HUMAN BODY

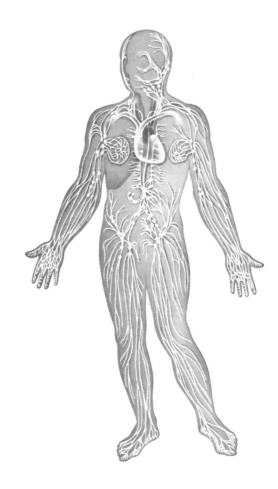

HUMAN BODY

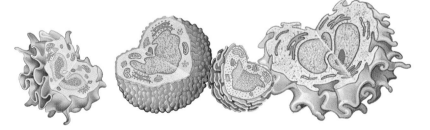

Oceana

A OCEANA BOOK

This book is produced by
Quantum Publishing Ltd.
6 Blundell Street
London N7 9BH

ISBN 0-681-78344-3

QUMTHB2

Manufactured in Singapore by
Universal Graphics Pte. Ltd.
Printed in China by
L. Rex Printing Co. Ltd.

CONTENTS

INTRODUCTION

As you look through the pages of this book, your body is busy at work—pumping blood, breathing, moving muscles, and sending messages to the brain. Housed in an ideal structure,

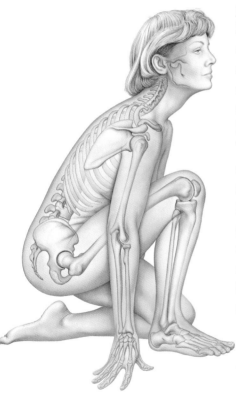

with custom-made parts and a cooperative labor force working 24 hours a day, our bodies are an amazing collaboration of design, engineering, and production.

The aim of this book is to present the human body and how it works, explaining the amazing work done behind the scenes by each and every part of the body, every second of every day, to ensure we can live our lives to the fullest. Written in everyday language and highlighted by detailed, full-color illustrations, this book will help the reader to gain a better understanding of the human body. For ease of reference, the book is divided into three sections: the body systems, the body regions, and the cycle of life.

The first section covers the main body systems, and

the role they play in our overall well-being.

In the second section, each region of the body is discussed, with informative illustrations of the organs, bones, muscles, nerves, and blood vessels that comprise each region, accompanied by concise text explaining how they work together in harmony.

The third section explains how the human body passes through many stages of development.

The *Human Body Pocket Guide* provides a useful resource for students of all ages, and is equally suitable as a family reference guide or for those who are simply curious to learn more about the fascinating workings of the human body.

Straightforward text and stunning illustrations ensure that learning about and understanding the human body—and how it works—is a rewarding experience.

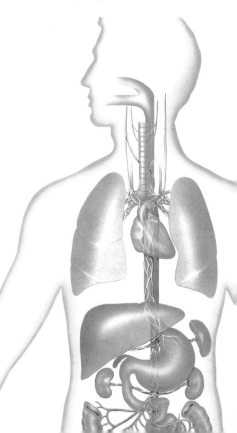

The
BODY SYSTEMS

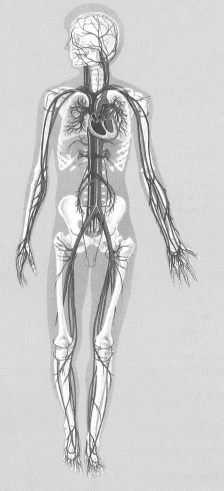

The human body is miraculous. Although we may marvel at the prowess of an Olympic athlete or the mind of the scientist, every human body is essentially the same. Details vary, but the nuts and bolts of every machine do not differ. All human nerves work in a similar way, everyone's heart is guided by the same rhythm. In this section of the book we look at body systems.

SKELETAL SYSTEM

Humans are vertebrates, creatures with backbones. They are reliant on a sturdy internal frame centered on a prominent spine. Bone is their being's cornerstone.

MUSCULAR SYSTEM

All motion follows mechanical principles, and the motion of the human body is no exception. Like all machines made by humankind, the body is a set of levers whose movements copy the geometry of classical mechanics.

CIRCULATORY SYSTEM

Blood is the stuff of life. Like most liquids, it takes on the form of its container—in our case, the human body. It dwells nowhere, has no "homebase"— although it belongs within a fraction of an inch of every cell in the body.

RESPIRATORY SYSTEM

Breathing is the most obvious and important of the many tasks accomplished by the respiratory system. But it is also involved in yawning, sneezing, coughing, sniffing—and hiccups.

Speech, from a whisper to a shout, and the subtle sense of smell, also depend on the movement of air in the respiratory passages.

NERVOUS SYSTEM

Silently and ceaselessly, at the height of human activity and during the depths of sleep, our nerves relentlessly transmit impulses to and from the brain.

DIGESTIVE SYSTEM

Most people enjoy eating, and food is a popular topic of conversation. On the outside, we call it food. It may be a crust of bread or a lavish 15-course banquet, but the digestion shows no favors. A healthy human eats about half a ton of it each year.

ENDOCRINE SYSTEM

The human body is a complex system of interrelated organs and tissues, all of which must work together if it is to function properly. Both monitoring and control have to take place without conscious intervention, and these are achieved by the endocrine system.

SKELETAL SYSTEM

When broken, it heals itself without scarring—a characteristic that sets it apart from nearly all other tissues in the body. Light yet strong, and seemingly simple in structure, the skeleton is nonetheless a masterpiece of architecture.

The human skeleton is built with the strength of an oak, yet it can also bend with a sapling's ease. It surrounds and protects the organs, supports the body and, bound by muscles, bestows the grace of movement. Ever building and breaking down, bone is dynamic tissue that forms in proportion to the task required of it. The bones in a ballerina's feet, a sculptor's hands, or a bricklayer's arm gain mass and alter in shape in response to the stresses their varied pursuits impose.

BONES

At birth, babies have about 350 individual bones. The adult skeleton totals 206 bones on average, resulting in a total of 26 movable parts in an adult. The smallest bones are in your ear. The largest bone, the femur, is found in your thigh.

A FLEXIBLE FRAMEWORK

Without the skeleton humans would find movement impossible. It provides a base for attaching muscles and the leverage to assist in their pulling. The skeleton protects the body's vital organs and provides life-giving substances—blood cells from red bone marrow and minerals from its bony storehouse. An ever-changing structure, it is also sensitive to the user's needs. The skeleton grows rapidly through childhood and adapts itself to our lifestyles, reinforcing areas where we, from sport, hobby, or occupation, add to the forces that already burden it.

The skeleton's most important function is that of support. Like beams that hold up buildings, the bones of the skeleton, assisted by muscles, hold the

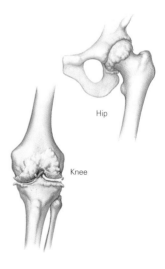

Hip

Knee

Skeletal System

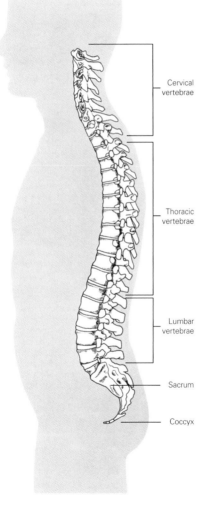

Cervical
vertebrae

Thoracic
vertebrae

Lumbar
vertebrae

Sacrum

Coccyx

body erect. But the body does not remain rock-still like a building. Where bone meets bone, joints provide the skeleton with flexibility and enable it to move.

Anatomists divide the skeleton into two broad categories—the axial and appendicular. The axial skeleton comprises the bones of the body's central core: the skull and spinal column which protect brain and spinal cord, and the rib cage, which shelters the heart and lungs. The appendicular skeleton includes all bones in the arms and legs, as well as the shoulder and pelvic girdles, which join the limbs to the axial skeleton.

A MULTIFACETED COLUMN

The spinal column is built of 33 small bones, the vertebrae. Seven cervical vertebrae in the neck are the smallest and allow the widest range of movement along the spine. Twelve heavier bones, the thoracic vertebrae, lie below them, forming the upper back. These bones also hold the 12 pairs of ribs in place. Special dish-shaped indentations called facets help anchor the ribs along each side of the vertebrae.

Five lumbar vertebrae in the small of the back are the largest bones of the spine and bear most of the body's weight. Tucked beneath the sacrum is the coccyx, a small, tapering bone built of four fused vertebrae. The coccyx is functionless and probably the vestige of a tail.

Skeletal System

THE SPINE

The spine is doubly curved like an elongated "S," and has evolved to support human beings in their unique upright posture. Its 35 vertebrae provide strength and flexibility.

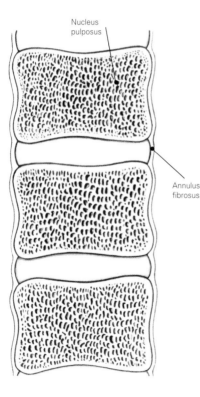

Nucleus pulposus

Annulus fibrosus

THE INVERTEBRAL DISKS

The vertebral column is made up of 26 bones that provide axial support to the trunk. The vertebral column provides protection to the spinal cord that runs through its central cavity. Between each vertebra is an intervertebral disk. The disks are filled with a gelatinous substance, called the nucleus pulposus, which provides cushioning to the spinal column. The annulus fibrosus is a fibrocartilageous ring that surrounds the nucleus pulposus, which keeps the nucleus pulposus intact when forces are applied to the spinal column. The intervertebral disks allow the vertebral column to be flexible and act as shock absorbers during everyday activities such as walking, running, or jumping.

RADICULOPATHY
This term comes from a combination of the Latin word "radix," which means the roots of a tree, and the Latin word "pathos," which means disease. Radiculopathy applies to a herniated intervertebral disk, or nucleus pulposus, when it has become displaced from its normal position in-between the vertebral bodies of the spine. This problem tends to occur most commonly in the cervical and lumbar spine.

Skeletal System

THE BUILDING BLOCKS OF THE SPINE

With the exception of the fused vertebrae in the sacrum and coccyx, and two highly specialized vertebrae at the top of the column, all the spine's bones are structurally similar. The main part of a vertebra, its body, is a flattened, oval block of bone. Short columns, the pedicles, arise from the back of the body. Small plates called laminae seal the opening between the two pedicles, creating a circular canal through which the spinal cord passes. Three winglike extensions crown the laminae. Spaced at 90-degree angles from each other, they anchor muscles and give the spine its knobby appearance under the skin. The articular processes, two smooth knobs on top of each vertebra and two beneath, link each vertebra with its neighbors.

The spine's topmost vertebra, the atlas, supports the skull. It does not have the solid body typical of the other vertebrae. It forms a bony ring with a large central opening. Rounded projections of the skull's occipital bone—the bone making up much of the skull's rugged base—fit into two large hollows atop the atlas. Tough ligaments bind skull to spine. The second neck vertebra, the axis, sends a bony projection into the base of the atlas. This articulation permits the atlas, holding the skull, to rotate at the top of the neck.

All along the spine, strong ligaments, muscles, and facets on the vertebrae bind the spine into a single column. Between adjacent vertebrae, there is a special cartilage padding called an intervertebral disk. Occupying about a quarter of an adult's spine, disks absorb shock and prevent the bones from grinding against each other. They allow motion between the vertebrae, bolstering the spine with added flexibility and strength.

The disks also play a role in shaping the spinal column, which rests at the body's vertical center of gravity. The spine crosses the gravity line in several places. The resulting curves provide much more stability and strength than a straight

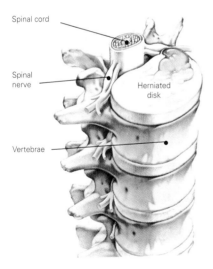

Spinal cord

Spinal nerve

Herniated disk

Vertebrae

ABOVE detail of the spinal cord in its place within the stack of the vertebrae in the spinal column.

Skeletal System

column could. To create curves, the disks subtly change shape, with portions narrowing so that the vertebrae do not sit directly on top of each other but rest at a slight angle. The overall effect is a series of alternating curves that run gracefully down the spine.

THE BONY CAGE OF THE CHEST

The spine is, in every sense, the backbone of the body and it directly or indirectly anchors all other bones. The ribs are joined to the thoracic vertebrae. Twelve pairs of resilient ribbons of bone springs from the sides of these vertebrae, their heads nestled in shallow facets. The upper seven pairs of ribs, the "true" ribs, arch around the body and attach to the sternum—the breastbone—via shafts of cartilage called costal cartilage.

The remaining five pairs of ribs are termed "false" because they join with the sternum indirectly. Costal cartilage links the upper three false ribs, which are connected, in turn, to the lowest pair of true ribs. The lower two pairs of false ribs "float," barely reaching around the sides of the body.

The ribs demonstrate the essence of the skeleton's protective yet flexible framework. The thorax, the bony cage they form, hugs the respiratory organs it houses. Yet the lungs must expand to do their work. Accordingly, the ribs swing outward and upward to accommodate the taking in of air. The sternum aids this process by maintaining a flexible space between two of its three main portions. Although shaped like the blade of a sword, this plate of bone acts as a shield, sheltering the heart and lungs.

HOW MANY RIBS?
Both men and woman have 12 pairs of ribs.

RIGHT The ribs provide protection to the vital organs of the chest cavity

Skeletal System

THE RIB CAGE

Twelve pairs of ribs encircle and protect the organs of the chest. Each pair is jointed by means of hollow facets to one of the thoracic vertebrae of the spine at the back and all, except the two lower pairs of short floating ribs, curve round to join with the sternum, or breastbone, at the front. The upper seven pairs of true ribs meet the sternum directly, while the next three pairs fuse with each other and the seventh rib. The joints themselves are made of rubbery cartilage, which allows the rib cage to move slightly with the movements of breathing.

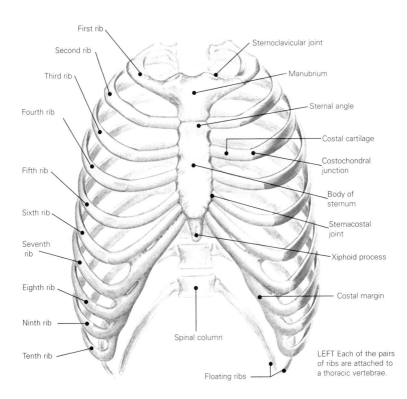

First rib

Second rib

Third rib

Fourth rib

Fifth rib

Sixth rib

Seventh rib

Eighth rib

Ninth rib

Tenth rib

Sternoclavicular joint

Manubrium

Sternal angle

Costal cartilage

Costochondral junction

Body of sternum

Sternacostal joint

Xiphoid process

Costal margin

Spinal column

Floating ribs

LEFT Each of the pairs of ribs are attached to a thoracic vertebrae.

Skeletal System

THE PROTECTIVE SKULL

Like the sternum, many bones in the body contribute to the protection of vital organs. The skull is made of 22 bones united in protecting the brain and major sense organs. Large at birth, compared to the rest of the body, the infant's skull is compressible. Soft spots lie between large pieces of bone that have not yet grown together. Passage through the birth canal usually squeezes the skull and elongates it slightly, but the head reassumes its natural shape within a few days.

The skull's soft spots, the fontanelles, are fibrous membranes that eventually harden and grow together until the bones meet and mesh much like the teeth of a zipper. The largest of the six main fontanelles is a diamond-shaped region near the center of the top of the skull. It is the last one to close, a process completed when a child is about 18 months old.

The cranium is built of eight bones that interlock at immovable joints called sutures. The frontal bone curves around the skull to create the forehead and roof of the orbits, bony sockets holding the eyes. Two temporal bones, forming the sides and a portion of the skull's floor, contain canals that lead to the middle and inner ears.

In the middle ear, just behind the eardrum, lie the smallest bones in the body. Each ear has a set of three bones: the malleus, incus, and stapes. The malleus picks up sound vibrations from the eardrum to which it is attached and passes them along to the incus and stapes. This relay method intensifies the vibrations; by the time they reach the stapes, their force has increased 20-fold.

The centerpiece of the cranium, the sphenoid bone, is bat-shaped with two large wings and two smaller "feet" extending from its central body. It forms much of the cranium's base and helps anchor most other cranial bones.

Fourteen bones combine to make the face. The main bones, the two maxillae, act together as a keystone for other facial bones. The maxillae articulate, or link, with all but the lower jaw. They form the entire upper jaw, part of the orbits, the nasal cavity, and the section of the roof of the mouth that contains the deep sockets for the upper teeth. At each side of the face, the ends of the maxillae swell

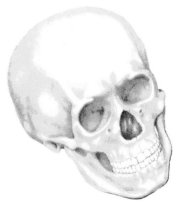

Skeletal System

forward and link with the small cheekbones.

The largest facial bone, and the only one in the skull with a movable joint, is the mandible, or lower jaw. It is a strong, curved bar of bone that holds the lower teeth. Arms of bone reach up its sides to articulate with the skull's temporal bones, creating the jaw's hinge joint. Tucked beneath the mandible is the U-shaped hyoid, the only bone in the body that

does not articulate with any other. It anchors muscles, particularly those of the tongue.

FETAL SKULL DEVELOPMENT
From its original template of cartilage, the skull gradually ossifies, and by birth the individual bones of the skull are formed, but not joined.

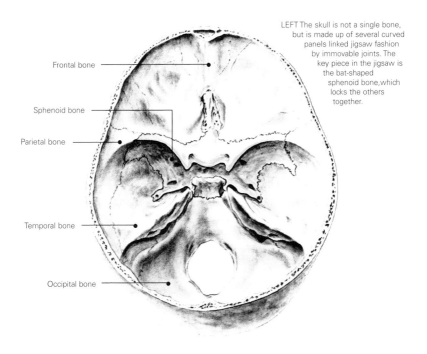

LEFT The skull is not a single bone, but is made up of several curved panels linked jigsaw fashion by immovable joints. The key piece in the jigsaw is the bat-shaped sphenoid bone, which locks the others together.

Frontal bone

Sphenoid bone

Parietal bone

Temporal bone

Occipital bone

Skeletal System

LEGS AND ARMS

The pelvic girdle is the bony construction that connects the legs to the rest of the skeleton. Charged with bearing the weight of much of the body, helping to hold it upright and move it forward, the pelvic girdle consists of two heavy, substantial hip bones joined by the sacrum at the back.

The leg bones, the body's supports, are much stronger than the bones of the arm, but are capable of much less movement. The largest bone in the body is the femur, or thigh bone. Long and elegantly proportioned, it is also the strongest and heaviest. At the top of its shaft, its smooth rounded surface snugly articulates with a socket in the outside of the hip bone.

The lower leg's main bone, the tibia, lies prominently at the front of the leg, where it is known as the shin. The lower leg's other bone, the fibula, resembles the jointed side of a safety pin and articulates with the tibia at its upper and lower ends. This slender, twisted bone acts mainly to anchor leg muscles. The two lumpy ankle bones are actually the prominent ends of each lower leg bone. The tibia surfaces at the inside of the ankle, the fibula on the outside.

The humerus is the long single bone in the upper arm. Like the femur, it has a smoothly rounded head which fits into a socket joint. The socket in the shoulder blade, or scapula, is shallow and much smaller than the head of the humerus.

FINGERS AND TOES
Toes have the same number of bones as the fingers and have the same anatomical name, phalanges. Also the big toe has two bones while the other toes have three.

This arrangement, which allows the arms to move at many angles away from the body but with little stability, makes the joint the most prone to dislocation.

The opposite end of the humerus meets two bones of the forearm, the ulna and the radius, to create the elbow joint. The predominant bone in the elbow is the topmost part of the ulna. The head of the radius is buttonlike, rounded at the edges and flat on top. This flat top glides against the surface of the humerus, and the rounded edges swivel against the ulna. Such a design greatly increases the range of movement of the arms.

The hand's great flexibility begins in the wrist, where eight carpal bones are arranged in two rows of four each. They look somewhat like small pebbles, yet their placement and articulation are precise. Beyond them, five metacarpal bones fan across the hand. Although largely alike they differ in length. Each of these bones articulates with one of the five digits. In each hand, there are 14 bones called phalanges, three in each of the fingers and two in the thumb.

The thumb is the most specialized of

Skeletal System

the digits and is the most important contributor to the hand's dexterity. This is due to the thumb's great individual movement. Its metacarpal bone is placed lower in the hand than those of the other digits and has a special joint, which links it to the wrist and allows the thumb to rotate freely across the surface of the palm. This permits the important hand movement known as "opposition," in which the thumb reaches across the palm to meet the other digits, fingertip to fingertip. Without this ability, the hand would be but a claw or an ungainly pair of forceps, with its precise movement and powerful grip severely limited.

The foot is hugely resilient to the great pressures and strains it encounters, and sacrifices dexterity to provide stability to the body. The ankle has only one bone less than the wrist, but the arrangement of its seven tarsal bones—different from that of the carpal bones in the wrist—increases their weight-bearing efficiency. When the body is standing, its entire weight funnels down through one of the tarsals, the talus. There, the weight divides evenly. Half travels to the calcaneus, the heel bone, while the other half channels to the five remaining tarsal bones and into the bones that make up the arch of the foot.

Slender metatarsal bones join in the central portion of the foot itself. These five bones articulate with the ankle's tarsals above and the bones of the toes below. They form the arch, the most important structural part of the foot. The foot disperses body weight through three arches—two over its length, one across its width. The arches also provide leverage for locomotion.

THE BONES IN THE TOES

Toes have the same number of bones as the fingers and have the same anatomical name, phalanges. They are also distributed similarly. The big toe has two bones and the others have three. Yet the bones of the toes have little in common with those of the fingers. Their stout shape enables them to bear weight. In a singular act of force, these bones help the foot push off with each stride as you walk or run. Firmly gripping the ground, they help us balance as we stand.

COMPACT BONE

Compact bone—one of the hardest tissues in the human body—is not completely solid, but is honeycombed with channels and spaces. Like the growth rings of a tree, bunches of columnar Haversian systems run the length of the bone. Each "ring" gets its strength and hardness from mineral crystals embedded in microscopic strands of the protein collagen. Cavities called lacunae contain osteocytes, which manufacture bone cells, and a central Haversian canal houses blood vessels carrying essential nutrients. The bone's outer surface is covered by the tough "skin" of the periosteum.

Skeletal System

THE STRENGTH OF BONE

By almost any measure, bone is among the strongest materials devised by nature. One cubic inch can withstand loads of at least 19,000 pounds—about the weight of five ordinary pickup trucks. This is roughly four times the strength of concrete. Indeed, bone's resistance to loads equals that of aluminum.

Bone seems even more remarkable considering its lightness. The skeleton accounts for only 14 percent of total body weight, or about 20 pounds. Steel bars of comparable size would weigh four or five times as much. Bone, ounce for ounce, is actually stronger than steel and reinforced concrete.

Unlike a steel shaft, which is useless when broken, bone is living tissue that repairs itself. Bone derives its strength by weaving protein and mineral into a resilient fabric. Each component enhances the strengths of the other. Nearly two-thirds of bone consists of various salts, mainly complex compounds of calcium and phosphorus. These salts form rod-shaped crystals, which lend bone hardness and rigidity. The remainder of bone is composed of collagen, an elastic protein which is converted to glue when bones are boiled. Collagen fibers are studded with the mineral crystals and wind around themselves like fibers in a rope. Remove the minerals and bone would become so rubbery that it could be tied in a knot, like a garden hose. Without collagen, it would be as brittle as glass.

THE THUMB
This digit is the most important contributor to the hand's dexterity due to its great individual movement which allows it to rotate freely across the surface of the palm.

Within bone tissue, the composite of collagen and mineral crystal forms elaborate structures. Compact bone is concentrated in long bones such as the femur. A cross-section of compact bone reveals an intricate pattern of concentric circles bunched together. Each self-contained cluster of circles, called a Haversian system, resembles the cut end of a tree trunk. Corresponding to tree rings are lamellae, layers of bony tissue made of crystal-studded collagen. At the heart of large systems lie Haversian canals, pipelines containing blood vessels, lymph vessels, nerve filaments, and delicate connective tissue.

Within the bone are small cavities in the lamellae called lacunae. A single cubic inch of compact bone contains more than four million lacunae. Each contains an osteocyte, or bone cell, which maintains mature bone tissue and supplies nutrients through minute channels. Radiating from each Haversian canal, the channels direct nutrients and remove waste from bone cells nestled in the lacunae.

The periosteum is a thin membrane which envelops the surface of compact bones like skin. It is laced with blood vessels, bringing nourishment to bones. If

Skeletal System

arteries that run through Haversian canals are blocked, vessels from the membrane can still supply local bone cells. Thin strands from the periosteum penetrate the bone and unite the two tissues. Other fibers entwine with tendons, securely anchoring muscle to bone.

Midway along the femur, the bone becomes hollow, yet does not lose mechanical strength. The resultant saving in weight is considerable. If the femur was a solid shaft, it would weigh 25 percent more. Nature has efficiently employed such hollow bones by filling them with marrow. In the center of ribs, vertebrae, pelvic, and skull bones lies red marrow, one of the body's most biologically active

LAYERS OF BONE
The different layers of bone comprise the inner core of bone marrow, surrounded by a layer of spongy bone, then compact bone, and an outer casing of periosteum.

tissues. Each minute, red marrow delivers millions of red cells, white cells, and platelets into the bloodstream.

Long bones, such as the femur, are filled with yellow marrow, consisting mainly of fat cells. Yellow marrow is a strategic store of energy reserves for occasions when the body's fat becomes depleted. Likewise, when blood becomes anemic through a lack of red blood cells, yellow marrow is quickly transformed into red marrow for the manufacture of red cells.

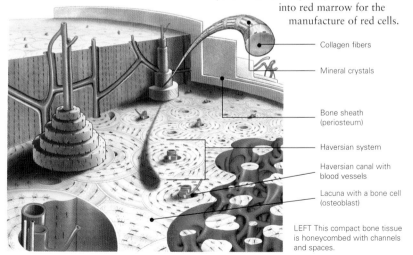

Collagen fibers

Mineral crystals

Bone sheath (periosteum)

Haversian system

Haversian canal with blood vessels

Lacuna with a bone cell (osteoblast)

LEFT This compact bone tissue is honeycombed with channels and spaces.

Skeletal System

MOVABLE JOINTS

Similar to the ones shown here, the skeleton incorporates several kinds of movable joints—a joint is where two bones meet and articulate. The articulation surfaces of these mobile joints, called synovial joints, are usually made smooth by a layer of slippery cartilage, which reduces friction. In larger joints lubrication is increased by synovial fluid. The bones are held in position by ligaments, which limit movements to the directions dictated by the joint's construction and the activating muscles' ability. The simplest type of joint is the hinge, as found in the elbows and the joints of the fingers and toes. It allows movement in only one direction. Gliding joints permit a wider range of mostly sideway movement and occur in the wrists, ankles, and spine. A saddle joint is yet more versatile, and gives the human

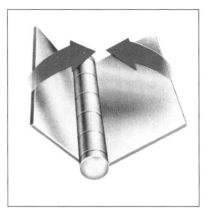

ABOVE The simplest type of joint—the hinge, as found in elbows and the joints of fingers and toes.

thumb its unique ability to "cross over" the palm. The widest range of movement is provided by a ball-and-socket joint, in shoulders and hips.

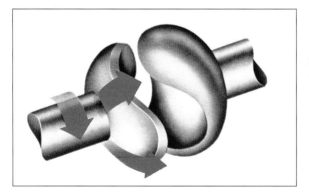

LEFT An example of a saddle joint. This can be found in the thumb, giving it a unique ability to "cross over" the palm.

Skeletal System

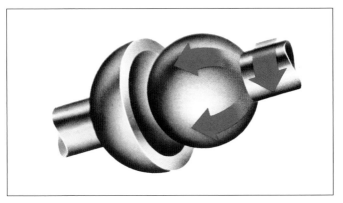

ABOVE This is a ball-and-socket joint which provides the widest range of movement in a joint. This joint is found in the shoulders and hips.

BELOW A gliding joint which permit a wide range of mostly sideway movement and are found in the wrists, ankles, and spine.

CARTILAGE
Cartilage is a resilient connective tissue containing cartilage cells and collagen in a matrix of glycoprotein. It provides a covering for joint surfaces, provides shock-absorbing qualities, and is a structural component of the skeleton where it joins two bone surfaces. There are three different types of cartilage and all have different physical properties: elastic cartilage, fibrocartilage, and hyaline cartilage.

Skeletal System

BONE BANKS

Repairing joints, implanting artificial joint replacements, and patching together the jaw of an accident victim are some of the marvels now performed by orthopedic surgeons. To do so they often need bone grafts, or transplants. Ideally such pieces of bone are autografts, taken from elsewhere in the patient's body. But autografts have limitations and are impossible in children (who cannot afford to "lose" any part of their growing bones).

The alternative is an allograft, using a piece of bone from a donor. To prevent problems with rejection or reabsorption, the graft is treated so that the recipient's body does not recognize it as "foreign."

THE EROSION OF CARTILAGE

Destruction of cartilage in an arthritic joint is initiated by the build-up of a deposit known as pannus. In it lurk leukocytes (white blood cells) and cellular debris. Other white cells called phagocytes, the scavengers of the body's immune system, arrive at the scene to mop up the remains. But they also release enzymes that attack the healthy tissues of

BELOW This knee is an example of how eroded cartilage first becomes scar tissue and then rough unyielding bone.

Cartilage

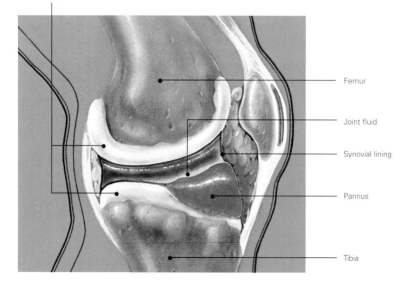

Femur

Joint fluid

Synovial lining

Pannus

Tibia

Skeletal System

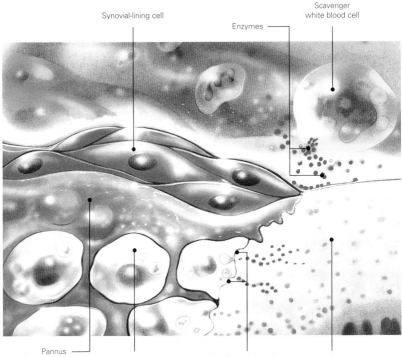

Synovial-lining cell

Enzymes

Scavenger
white blood cell

Pannus

White blood cell

Cartilage destruction

Cartilage

ABOVE White cells, called phagocytes, release
enzymes that attack healthy tissue of the joint lining.

the joint lining. A commonly affected
joint is the knee in which eroded cartilage
first becomes scar tissue and then rough
and unyielding bone. The progress of
rheumatoid arthritis is now well known,
but despite intensive research the cause
remains a mystery.

THE BONE BANK
Pieces of donor bone, taken from
cadavers, are freeze-dried and stored at
room temperature, or deep-frozen and
stored in a bone bank.

Skeletal System

WHERE BONE MEETS BONE

The need for strength makes bones rigid. If the skeleton were cast as one solid bone, movement would be impossible. In all vertebrates, nature has solved this problem by dividing the skeleton into many bones and creating joints where bones intersect. Joints come in an array of designs, each custom-built for the limb it serves. Lashed together by fibers of collagen called ligaments, and continuously lubricated to offset friction, joints permit movement.

Some joints, such as those binding the skull's protective plates of bone, allow no movement. Others permit only limited movement: the joints between the spinal vertebrae are united by disks of cartilage and other tissues, which allow some movement in several directions.

Most joints offer far greater play of movement. Connections called synovial joints are sturdy enough to hold the skeleton together while permitting a range of movements. The structure of synovial joints helps transmit power and motion between the bones. The ends of bones at synovial joints are coated with a tough articular cartilage, which reduces friction and cushions the joint against jolts. Lying between the bones is a narrow space known as the joint cavity, which offers freedom of movement. Ligaments bind the bones, prevent dislocations, and limit the joint's mobility.

Synovial joints are fulcrums, the bones they connect are levers, and the muscles

WEIGHT-BEARING JOINTS
Weight-bearing joints, such as those of the hips, knees, and spine, rely on their cartilage linings to operate smoothly. Breakdown of the cartilage, as occurs in osteoarthritis, leads to friction, erosion of the joint, and pain. The joint becomes inflamed and stiff, and in severe cases may seize up altogether and cause crippling disability. Osteoarthritis in the joints of the hand results in nodular swelling of the knuckles.

attached to them apply force. The joint between the skull and the atlas vertebrae of the spine is the fulcrum across which muscles lift the head. Nodding one's head would be impossible without it. When you lift a book, the elbow joint is the fulcrum across which the biceps muscle performs the work.

A sleevelike extension of the periosteum envelops every synovial joint. This joint capsule permits movement within the joint, but it also has sufficient tensile strength to prevent dislocation. Lining the joint capsule, a fine membrane secretes the lubricant synovial fluid. The fluid oils the joint while cells within it remove microorganisms and debris.

Bursae are closed sacs with a lining not unlike that of synovial joints. Bursal walls secrete a substance similar to, but less viscous than, synovial fluid. The sacs help one musculoskeletal structure to

Skeletal System

glide over another, for example skin or tendon over bone. The knee has at least 12 such sacs.

Each type of synovial joint is precisely designed for a specific movement. The most freely moving are ball-and-socket joints in which the hemispherical head of one bone lodges in the hollow cavity of another. In the shoulder joint, the humerus bone of the upper arm fits into the socket of the shoulder blade. Because the socket is shallow and the joint loose, the shoulder is the body's most mobile joint.

The ball-and-socket of the hip joint is less mobile than the shoulder, but more stable. The ball of the femur's head fits tightly into a deep socket in the hip bone. A rim of cartilage lining the socket helps grip the femur firmly; the ligament binding the two bones is among the strongest in the human body.

The saddle joint connecting thumb to hand permits movement in two directions. Hinge joints, such as elbow and finger joints, are less mobile and allow movement in only one direction, like the hinge on a door. A pivot joint near the top of the spine allows the head to swivel and bend. Another pivot joint, in the forearm, allows the wrist to twist.

The hinge joint of the knee, the body's largest joint, is unusual because it can swivel on its axis, allowing the foot to turn from side to side. Thus the knee is constantly rolling and gliding during walking. Two crescent-shaped wedges of cartilage called menisci permit the joint to

BONES STOP GROWING

In most people bones continue to grow and develop until about the age of 25, by which time bone has replaced cartilage, ending further growth.

glide easily because of their slippery surfaces. Menisci are also shock absorbers, cushioning the blows the joint endures in everyday use. They are the knee's weak link; torn menisci account for 90 percent of all knee surgery.

GROWTH AND RENEWAL OF BONE

Bone grows by slow accretion, adding layer upon layer until adulthood. It changes constantly, growing and repairing itself like the rest of the body. Bones develop and harden through the process of ossification, which begins around the fifth week of pregnancy. Most of the developing bones in a human fetus are first evident in a flexible skeleton of cartilage, a tough gristlelike substance the color of milky glass. For many people, ossification continues until about the age of 25, by which time bone has replaced cartilage, ending further growth.

Skeletal System

TEETH

Although composed mainly of calcium-containing minerals, teeth are harder than bones. The illustrations below represent the upper and lower halves of the right side of the jaw.

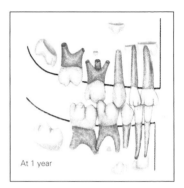

At 1 year

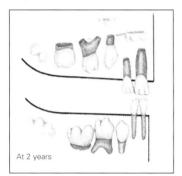

At 2 years

These diagrams show the order of eruption of a child's 20 milk, or first, teeth (blue), which are followed by the 32 permanent, or second teeth (yellow). The third molars, commonly called wisdom teeth, may never erupt.

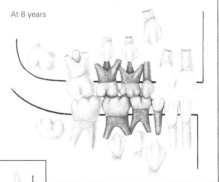

At 8 years

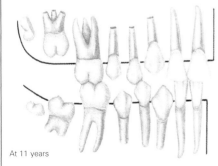

At 11 years

THE STRENGTH OF TEETH
Teeth are even harder than bones, but like them are composed mainly of calcium-containing minerals. The outer visible part, the crown, is composed of enamel—the hardest substance in the body.

Skeletal System

DISEASES OF THE BONE

Skeletal growth is regulated by secretions from the pituitary, thyroid, adrenal, and sex glands. Proper nutrition strengthens bone, particularly when adequate amounts of vitamins A, C, and D are present. Pressure on the ends of bones, whether from cartilage, muscular development, or injury, can also influence growth.

The most prevalent reminder of the skeleton's ultimate tendency to fail mechanically is arthritis, a disease with a long legacy. Arthritis is broadly defined as the chronic inflammation of the joints and their subsequent degeneration. More than 100 varieties of arthritis are known, but these can be grouped into three categories: gout, osteoarthritis, and rheumatoid arthritis.

Gout is the best understood and most treatable of the three. It is brought on by a rich diet high in purines, compounds that yield uric acid when metabolized by the body. This acid crystallizes if it accumulates in the blood in sufficient quantity. The crystals then migrate to joints, especially in the big toe, where they cause inflammation and pain.

Osteoarthritis is believed to be essentially mechanical in nature, because it comes about mainly from a lifetime's wear and tear. The disease turns up mainly in the load-bearing joints—hips, knees, and spine—and in the hands. Curiously it afflicts women twice as often as men. It occurs when cartilage, serving as a shock absorber between bones,

begins to break down. Bone starts to rub against bone, causing severe pain and reduced mobility. The regenerative chemistry of bone tissue alters, unable to prevent the soft edges of the worn cartilage hardening into bone spurs.

In rheumatoid arthritis, the synovial membrane lining a joint becomes inflamed. Malfunctioning white blood cells and joint tissue cells form a deposit called pannus on the joint surface. Enzymes released by the inflammation mysteriously turn on the joint and start to digest it. Scar tissue forms between adjacent bones and hardens into bone, which fuses the joint. Rheumatoid arthritis may be triggered by a volatile mix of a yet-to-be identified virus and a faulty immune response. In this scenario, white blood cells alerted by the infection mistake the joint for the enemy and attack it.

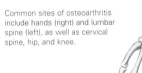
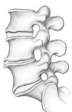

Common sites of osteoarthritis include hands (right) and lumbar spine (left), as well as cervical spine, hip, and knee.

MUSCULAR SYSTEM

The body is made up of a set of levers that are powered by muscles, the elegant, efficient body components whose actions are as simple as their character is complex. The day-to-day operation of muscles, which we all take for granted, belies a division of labor so elaborate that scientists have only just begun to unravel it.

Each of more than 600 muscles is served by nerves. Linking muscles to the brain and spinal cord, a network of nerve circuits carries signals that direct the ebb and flow of muscular energy. Many muscles must work together to perform even the simplest tasks. And much muscular activity occurs outside the realm of the conscious mind as the body, through the neuromuscular network, manages its own motion.

Muscles move—and by their motion we move. Yet despite the variety of actions we are capable of performing, muscle itself moves only by becoming shorter—it pulls but it cannot push. And while whole muscles can be seen to contract in this way, they consist, in reality, of millions of tiny, finely tuned protein filaments working together in superb synchrony.

Bodily functions demand that muscles accomplish different chores. Accordingly, there are three distinct types. Cardiac muscle, found only in the heart, powers lifelong pumping. Smooth muscle surrounds or is part of internal organs and the blood vessels that fuel them. Both are described as involuntary, since they are not usually under our conscious control. But the word "muscle" brings something else to mind. What we will

into action, what gives the body form, what aches after a ten-mile hike is skeletal muscle, muscle that carries out voluntary or conscious movements. Skeletal muscle is anchored to bones and pulls them to initiate movement. Accounting for 23 percent of body weight in women and 40 percent in men, skeletal muscle is the body's most abundant tissue.

Weightlifters, baseball pitchers, and sprinters exhibit some of the many remarkable displays of force the human body is capable of performing. Yet such force is possible only through the arrangement of the muscles, bones, and joints that make up the body's lever systems. Bones act as levers, while joints perform as living fulcrums, the pivots about which vital powers are exercised. Muscle, attached to bones by tendons and other connective tissue, exerts force by converting chemical energy into tension and contraction. When a muscle contracts it shortens, in many cases pulling bone

THE LARGEST MUSCLE

The largest muscle is the gluteus maximus, in the buttock, which is used when we jump up and down and leap forward.

Muscular System

like a lever across its hinge. This movement depends on many factors, such as the stabilizing influences of other muscles that contract at the same time.

Muscular action also involves the senses. Progressive movements constituting behavior, from a smile to a ballerina's arabesque, are possible only through incessant communication between senses and muscles. This constant sensory-motor integration occurs between the spinal cord and the part of the brain that coordinates sensation and motor response.

There are more than 600 voluntary muscles in the body, which together account for about 40 percent of a man's weight. The strongest are the major skeletal muscles shown here. Their contraction causes all body movements, from bending and running to lifting and gripping. Even subtle changes of facial expression are brought about by delicate adjustment of the small muscles in the face. Strong non-elastic tendons join muscles to bones, where they exert their pull. The fingers, for example, are bent and straightened by muscles in the forearm connected to the finger bones by long tendons.

HOW DO THE FINGERS MOVE?
Strong non-elastic tendons join muscles to bones, where they exert their pull. Therefore the fingers are bent and straightened by muscles in the forearm connected to the finger bones by long tendons.

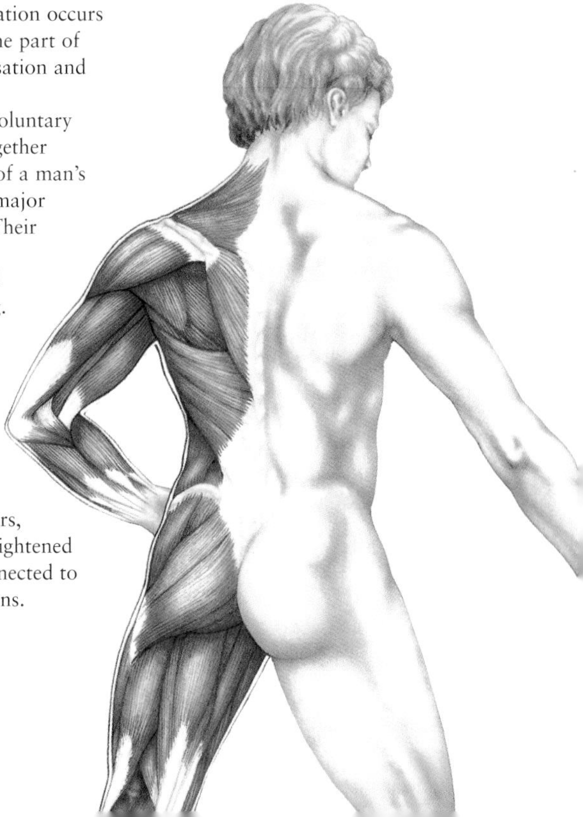

Muscular System

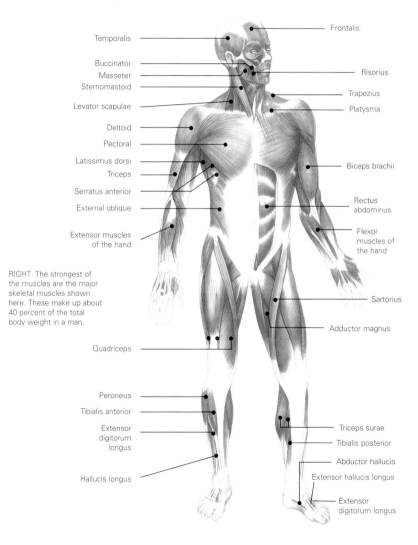

Temporalis

Frontalis

Buccinator

Masseter

Sternomastoid

Risorius

Levator scapulae

Trapezius

Platysma

Deltoid

Pectoral

Latissimus dorsi

Triceps

Biceps brachii

Serratus anterior

External oblique

Rectus abdominus

Extensor muscles of the hand

Flexor muscles of the hand

RIGHT The strongest of the muscles are the major skeletal muscles shown here. These make up about 40 percent of the total body weight in a man.

Sartorius

Adductor magnus

Quadriceps

Peroneus

Tibialis anterior

Extensor digitorum longus

Triceps surae

Tibialis posterior

Abductor hallucis

Extensor hallucis longus

Hallucis longus

Extensor digitorum longus

Muscular System

A CONSTANT SUPPLY OF OXYGEN

Muscles need a constant supply of oxygen in order to work. This vital gas is carried in the bloodstream along arteries, whose small branches encircle each bundle of muscle fibers. Even individual fibers have their own network of capillaries, in which gas exchange takes place: oxygen is supplied to the fiber in exchange for carbon dioxide. This waste gas is carried away in the blood along veins, which transport it to the lungs for disposal.

SOME OF THE STRONGEST MUSCLES
The strongest are the major skeletal muscles. Their contraction causes all body movements, from bending and running to lifting and gripping. Even subtle changes of facial expression are brought about by delicate adjustment of the small muscles in the face.

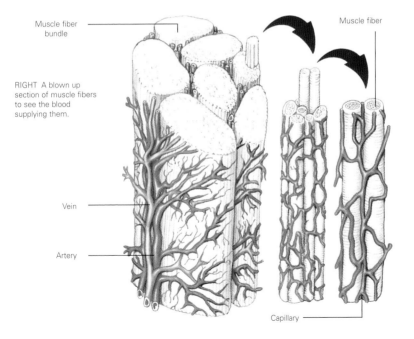

Muscle fiber bundle

Muscle fiber

RIGHT A blown up section of muscle fibers to see the blood supplying them.

Vein

Artery

Capillary

Muscular System

MUSCLES IN MOTION

"Motion is the cause of all life," wrote Leonardo da Vinci in his notebooks. Indeed, the movements of the human body, and many of its vital internal functions, are guided and coordinated by muscles of varying shapes and sizes. Yet before they can move the body, the muscles must first be able to maintain its characteristic upright posture against the collapsing force of gravity. Just standing still and upright depends on the contraction of a whole range of different muscles.

Control of the muscles necessary to standing and moving is not something babies are born with. But gradually, over the early years of life, they learn the coordination and control that will, barring accidents, disease, and the ravages of old age, last them a lifetime. Babies learn to control their muscles from the head down: muscles of the neck, followed by those of the shoulders and arms, then the body. Only with mastery over muscles of the pelvis and legs are standing and walking possible.

A FLUENT SYNCHRONY

The apparently simple act of walking thus takes time to perfect. It also involves specific actions by many different muscles. In the human body, every muscle has a particular function, but each works in a fluent synchrony with others to achieve its role. Muscles called prime movers, such as the deltoid muscle in the shoulder, are powerful imitators of force. The triangular deltoid is a weightlifter's prime mover when he raises a heavy load above his head.

Because their contraction results in movements, such muscles are called agonists. Their action is opposed by muscles called antagonists: any muscle that extends a limb is acting antagonistically to a muscle that flexes or bends it. The triceps in the upper arm, for example, is antagonistic to the biceps when the arm is bent at the elbow.

Such paired muscles as the biceps and triceps faithfully alternate roles as agonists and antagonists in perfect collaboration, making possible the cooperation necessary to smooth and efficient muscular effort. When a prime mover contracts, the tension in its antagonist slackens and stabilizes

MUSCLES IN MOTION
Oddly, babies often run before they can walk. Typically, babies on the verge of true walking stand with the feet wide apart, ensuring a wide base of support for the body. They lean forward, thus advancing the center of gravity and, in a natural though erratic sequence of steps, can maintain an upright posture for a few seconds as they scamper to a person or piece of furniture for support.

Muscular System

movement. Assistant movers, muscles that contribute to a specific movement, often assist the prime mover in making muscular expression possible. The hamstring muscle of the thigh is the prime mover in bending the knee, the sartorius is its assistant and the quadriceps its antagonist.

The role of fixator or stabilizer muscle is to hold a bone or other body part steady, so providing a firm foundation upon which active muscles can pull.

MUSCLE FIBER
The filaments in muscle fibers are made of two kinds of proteins, actin and myosin, linked by cross bridges. The thin filaments are formed of actin, and the thick ones of myosin.

In weightlifting, for example, abdominal muscles contract to prevent sagging of the hips and trunk and to enable the transfer of momentum from the body to the weight as the weightlifter stands up.

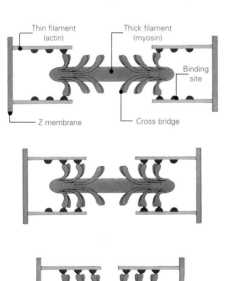

Thin filament (actin)

Thick filament (myosin)

Binding site

Z membrane

Cross bridge

LEFT As muscle contracts, and the filaments slide over each other, myosin "fingers" latch on to a series of binding sites on the thin actin filaments to form more cross bridges and create a much stronger and denser structure.

Muscular System

FILAMENTS IN MUSCLE FIBER

The filaments in muscle fibers are made of two kinds of proteins, actin and myosin, linked by cross bridges. The thin filaments are formed of actin, and the thick ones of myosin. Resting muscle has an open structure with few links between the two types of filaments. As muscle contracts, and the filaments slide over each other, myosin "fingers" latch on to a series of binding sites on the thin actin filaments to form more cross bridges and create a much stronger and denser structure.

FAST-TWITCH FIBERS

Skeletal muscles are composed of two kinds of fiber in proportions that vary according to function. Fast-twitch fibers provide strength and power; they are the "white meat" of muscle. They contract quickly, yielding short bursts of energy and are recruited most heavily for brief,

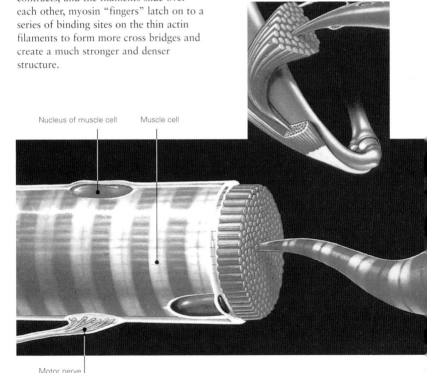

Nucleus of muscle cell   Muscle cell

Motor nerve

Muscular System

intense exercises—sprinting, weightlifting, putting the shot, or swinging a golf club. But these muscles are quick to exhaust—cramp sets in as they become vulnerable to build-up of lactic acid, a by-product of their own metabolism.

SLOW-TWITCH FIBERS

Slow-twitch fibers produce a steadier tug and are the cords of endurance, tiring only when their fuel supplies are consumed. Slightly smaller than fast-twitch fibers, and containing fewer nerve endings, they draw more oxygen from the

blood. They are the dark meat of muscle, their color imparted by their abundant blood supply. Exercises requiring enormous stamina—long-distance running, swimming, or cycling—rely on the sturdy strength of slow-twitch fibers.

LEFT and BELOW Muscles look solid but are really made up of bundles of tiny fibers. Each fiber is a muscle cell, which has a fibrous microstructure of its own, consisting of thick and thin filaments in units

called sacromeres. Signals arriving along a motor nerve make the filaments slide over each other, causing the whole muscle to contract.

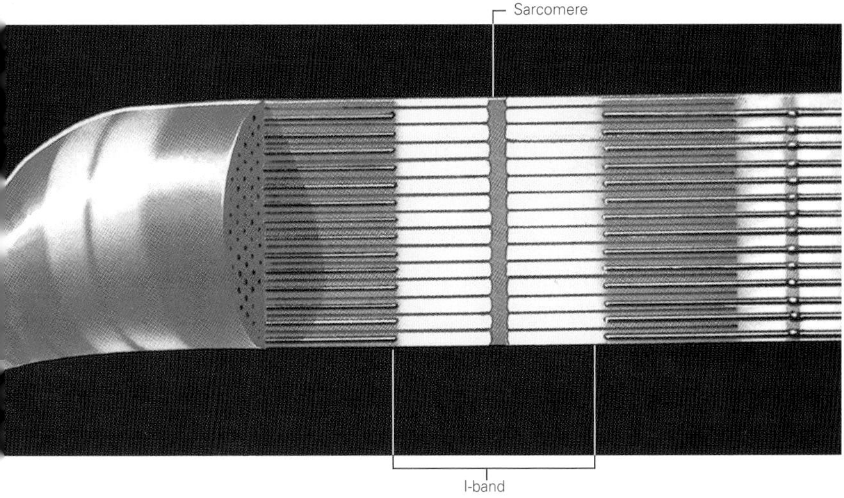

Sarcomere

I-band

Muscular System

SYSTEMS OF LEVERS

To keep the body in action, muscles work in concert with bones and joints in lever systems that obey the basic physical rules of motion and energy. To return to the business of walking: placing one foot in front of the other, pace by pace, is a process in which balance is continually lost and regained as each step establishes a new base. The legs and arms act as lever systems, pivoting on fulcrums of ankles, shoulders, and hips. The legs generate a rotary motion that propels the body forward and momentarily upsets the body's equilibrium. By synchronizing the movements of the arms and legs, the muscles ensure that balance is restored before the body becomes unstable and falls.

The impact of the muscle groups working together is called power. For more power, the nervous system must animate more muscle fibers. Muscles exert this power upon the points where they are attached to bones (their points of insertion). Bones linked together by joints form lever systems: the arm lever bends, for example, at the elbow's hinge joint and swings or rotates at the shoulder's ball-and-socket joint.

In all, there are only three types of lever system in the body. Each lever has a load arm—the zone between the load to be managed and the fulcrum—and a power arm, which is the region between the fulcrum and the muscle. This power arm is nearly always shorter than the load arm.

The muscles that act on the power

THE LONGEST MUSCLE
The longest muscle is the sartorius, which runs from the side of the waist, diagonally down across the front of the thigh to the inside of the knee.

arm develop their greatest power by shortening slowly while, at the same time, developing high tension. The force is amplified in the lever system producing rapid movement of the load arm. Called first-class levers, they have a fulcrum sited between the force and the resistance. The throwing action of a baseball pitcher uses this kind of lever—short, relatively slow movements of arm muscles produce rapid movements of the hand.

In second-class levers, the resistance is between the fulcrum and the muscular force that is pitted against it. Here, power rules rather than speed. A ballet dancer, on half-point in an arabesque, balances the body weight with an upward pulling calf muscle, using the toes as a fulcrum.

In third-class levers, the force is applied between the fulcrum and the resistance. The faster the muscle contracts, the greater is the force that can be applied at the end of the moving lever. The slower the contraction, the less energy is required to produce it. The action of the biceps as it exerts an upward pulling force on the forearm when the elbow is bent is an example of such a lever.

In the human body, speed and range of motion prevail over force. The body is

Muscular System

best at tasks involving fast or delicate movement, and the movement of light objects. When great force is demanded, as during heavy work, the body usually proves inadequate. To compensate for this, the human mind has invented machines and tools, such as the crowbar, which extend muscular leverage to gain a force advantage.

FIBROUS BUNDLES

The structure of muscles is designed to enable them to contract and relax, and they provide the body with a lifetime of movement. All muscles are made up of fibers, but each of the three types of muscle has a different microscopic structure related to its exact role in the body machine.

In skeletal muscle the fibers are elongated cylinders. Each fiber contains several nuclei (the sites where genetic material is housed). These nuclei originally belonged to myoblasts, smaller "pre-muscle" cells that merged together before birth. Because skeletal muscle fibers are much larger than cardiac or smooth muscle fibers, many are visible to the naked eye. Some like those in the sartorius muscle of the thigh, are more than 12 inches long.

Individual skeletal muscle fibers can extend the whole length of a muscle. Usually, however, one end attaches to tendon, the tough tissue that binds muscle to bone, and the other attaches to con-nective tissue in the muscle itself. The

LEVERS FUNCTION IN UNISON
Physical work most often involves the use of more than one of the body's lever systems. Even when movement of a single lever does take place, many other parts of the body must be held in place.

firm, white tendons form a kind of core for a muscle by extending far inside it and emerging at the muscle ends to link to bone.

Fibers of muscle and tendon are completely different materials, however, and do not merge. Instead, connective tissue extending from the tendon links with the end of the muscle fiber.

Surrounding each muscle fiber is the endomysium, a thin sheath of connective tissue. Another sheath, the internal perimysium, bundles the individual fibers into fasciculi, groups of about 12 fibers. These groups are themselves bound together by another layer of connective tissue called external perimysium or epimysium. It is this final grouping of bundles that is commonly referred to as muscle.

Muscles enlarge with use because exercise, especially lifting weights, stimulates the production of greater amounts of actin and myosin, the special proteins that muscle contains, so expanding the fibers. Primed for heavy work, such muscles become well defined under the skin. But muscle enlargement, or hypertrophy, is not as great in women as in men because it is partly regulated by the male sex hormone testosterone.

Muscular System

THE STRANDS OF MUSCLES

The dynamics of muscle are locked in its basic component, the fiber. Each individual fiber is surrounded by a thin plasma membrane, the sarcolemma. Some 80 percent of the fiber's volume is filled with tiny fibrils, known as myofibrils—from several hundred to several thousand, depending on the width of the muscle fiber. The remainder of the fiber is filled with a jellylike intracellular sarcoplasm, the many nuclei and other constituents of any typical body cell, such as mitochondria in which energy-producing reactions take place.

Thin dark bands called Z-membranes separate the myofibrils into cylindrical compartments called sarcomeres—the basic units of a muscle's contraction. A prominent dark strip, the A-band, occupies the center of each sarcomere, the space between each A-band is taken up by the paler I-band. This exact pattern is repeated down the length of every myofibril.

FIXED AT BIRTH
The number of fibers in any given muscle is fixed at birth; damaged fibers can never be replaced, even by a healthy body. A weightlifter has no more fibers than the proverbial 97-pound weakling, but he does have larger muscle fibers and muscles with more connective tissue.

Just as a single fiber contains many myofibrils, so each myofibril contains many smaller filaments arranged in a repeating pattern along the fibril's length. There are thick filaments made up of the protein myosin and thin filaments composed of the protein actin. The arrangement of the filaments gives rise to the striations. The thick filaments form the dark regions of the sarcomere; the thin filaments the light areas. Thus, the dark A-band in the center of the sarcomere consists mostly of myosin filaments anchored at the M-line, the center of the A-band.

The thin actin filaments, anchored to the Z-membranes, form the light I-bands. The darkest striations occur when myosin and actin filaments overlap. The A-band is darkest because it normally includes both types of filaments. However, the actin filaments do not quite stretch into the center of the sarcomere, which is why the H-zone appears lighter.

THE POWERS OF CONTRACTION

The term "contraction" does not always refer to the shortening of a muscle. Technically, it refers only to the development of tension within a muscle. There are two major types of contraction. A contraction in which the muscle develops tension but does not shorten is isometric, and one in which the muscle shortens but retains constant tension is said to be isotonic. Both are determined by the amount of resistance the muscle

Muscular System

meets as it contracts.

When a fiber contracts, the length of the dark A-bands remains constant, but the two pale regions (the I-band and the H-zone) shorten. The protein filaments themselves do not shorten but slide past one another, like lines of soldiers marching in opposite directions.

The muscle machine swings into motion only when it receives impulses

THE AMOUNT OF MUSCLE FIBERS
A single motor unit in the tiny muscles of the eye may contain only three muscle fibers, whereas those in the relatively slow-moving large muscles such as the calf muscles house a thousand or more.

from the central nervous system which tell it to do so. The nerves terminate near the muscle fiber's delicate membrane, where they release transmitter chemicals. These neurotransmitters initiate a wave of electrical activity that spreads through the whole fiber. This causes the fiber's membrane to release calcium ions, electrically charged calcium atoms that spark the mechanical process of contraction.

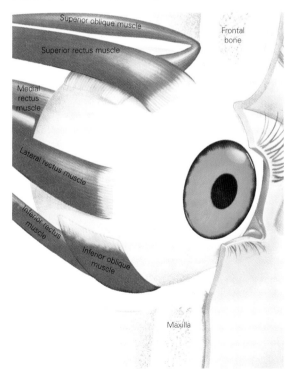

Superior oblique muscle

Superior rectus muscle

Frontal bone

Medial rectus muscle

Lateral rectus muscle

Inferior rectus muscle

Inferior oblique muscle

Maxilla

LEFT The six muscles of the eye work in unison to offer a wide field of vision, by providing up, down, left, and right movement.

Muscular System

CROSS BRIDGES BETWEEN FILAMENTS

Emerging from the myosin filaments are pairs of rounded buds. Each pair forms the head of a single myosin molecule. These highly-ordered projections called cross bridges, form the central part of the muscle's contractile mechanism and take their name from that role.

Cross bridges are crowned with a remarkable substance, adenosine triphosphate (ATP). ATP is an organic compound, and the main source of life's energy derived from the food that we eat. The high-energy ATP is transformed into two low-energy products: adenosine diphosphate (ADP)—which has two phosphates incorporated in it instead of ATP's three—and the inorganic phosphate that splits away. The energy lost in the split is then available for use in the body's metabolism.

STRIATED MUSCLE

Skeletal muscle is often referred to as striated muscle for the simple reason that it appears striped or banded when viewed through a microscope. The striations arise not from the surface of the fiber but from its many myofibrils, each of which is banded. The parallel arrangement of the myofibrils within the fiber give its characteristic appearance.

ATP and the myosin molecules are said to have a high affinity for each other—so much so that in normal muscle almost every myosin head has ATP resting on it. Each of the head's two buds has a different function. One contains the enzyme adenosine triphosphatase (ATPase), which has the ability to split ATP molecules and liberate energy as a result. The other portion of the head binds to the actin filaments.

As the protein tropomyosin shifts away from the binding sites on the actin filaments, the myosin arms are able to link with actin. The myosin ATPase then splits ATP and its energy fuels rapid contraction of the muscle machinery by throwing the cross bridges into action. Without ATP the actin and myosin filaments remain locked together, and the muscle is unable to contract. This is what causes rigor mortis, the stiffening of muscles shortly after death. Dead muscle cells have no ATP.

Contractions normally shorten a sarcomere by a fifth or more, cross bridges making and breaking links with the actin filament in repeated cycles. Like a team working hand over hand to pull a rope, cross bridges repeat this action until the contraction is complete.

The more cross bridges there are the more powerful is the fiber's contraction. When a muscle is stretched too much there is little or no overlap between the thick and thin filaments, so cross bridges cannot make enough connections to create tension. On the other hand, with

Muscular System

too much overlap themselves, thin filaments on either side of the sarcomere begin to overlap themselves, interfering with the action of the cross bridges. And, in a greatly shortened fiber, the thick filaments are compressed between the two Z-membranes, making further shortening difficult. For this reason, a muscle has one particular length at which it contracts most efficiently. Normally, resting muscle is quite near this length. The greatest tension is produced when a muscle is stretched slightly beyond this length, and it contracts more forcefully.

LIFTING A HEAVY WEIGHT

A person trying to lift too many books in a box strains against the weight. His arm muscles develop tension but do not shorten because the amount of resistance offered by the box is greater than the muscle's tension. But when he lightens the load, the working muscles shorten as they contract. This is an isotonic contraction. Because his muscles shorten in overcoming the resistance, the isotonic contraction is said to be concentric. If he wants to put the box down on a table, he must gradually extend his arms. To do so, the biceps in the upper arms lengthen, maintaining tension to counter the weight of the box. The biceps approach, but do not reach, their resting state. In this case the isotonic contraction is described as eccentric because the muscles lengthen as they act to maintain tension.

POWER OF CONTRACTION
Muscles must counteract gravity. Although they may relax fully, they also remain alert in case they are called into play to prevent the body from falling. Scientists use the word "tone" to describe this constant state of readiness, which is determined by a healthy nervous system. Without tone, jaws would hang open and muscles could not support parts of the body.

MOVEMENT AND THE BRAIN

From all parts of the body nerves known as sensory neurons carry impulses to the brain or spinal cord, carrying messages about the state of affairs in every body part, including the muscles. Conversely, motor neurons transmit impulses to muscles, often via intermediate connections or interneurons in the spinal cord. Motor impulses travel to the muscle fibers where they trigger the release of the neurotransmitter substance acetylcholine. This crosses the gap or synapse at the junction between nerve and muscle, setting off a chain of events that ends in contraction. Within a second, millions of impulses reach the spinal cord, others from special sense organs located in joints, ligaments, tendons, and muscles themselves.

Muscular System

THE CEREBRAL CORTEX

The seeds of movement, however, are sown by the brain, in its primary cortex, a region of the brain's wrinkled surface spanning both cerebral hemispheres. This motor cortex lies in a ridge just in front of the central sulcus, a deep furrow running vertically along each of the cerebral hemispheres. And a patch of cortex directly in front of the primary area also houses neurons involved in movement. This secondary motor area is thought to be crucial to speech and intricately coordinated movements such as those the hands perform.

HAND MUSCLES
We perform many delicate movements with our hands, and these are driven by impulses from roughly 200,000 pyramidal neurons.

BELOW Muscular movement is coordinated by various areas of the brain, principally in its outer layer, the cerebral cortex. The primary motor area, a band across the center of the cortex, controls gross movements; but when fine, intricate movements are required the secondary motor area comes into play. Input from the various senses passes first to the somatic sensory area, which supplies feedback vital to any voluntary muscular action, before reaching the motor areas.

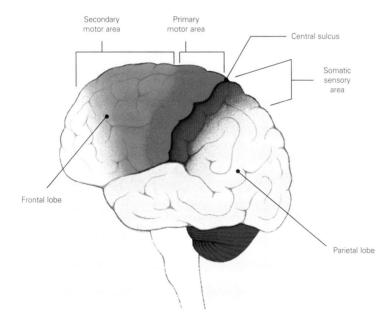

Secondary motor area

Primary motor area

Central sulcus

Somatic sensory area

Frontal lobe

Parietal lobe

Muscular System

Electrical impulses from many regions of the brain feed into the motor areas. Before the nervous system can orchestrate a coordinated movement, it must collect and integrate all sensory messages. In the somatic sensory cortex lying on the other side of the central sulcus the brain registers sensations. Interplay between the senses and movement is continuous and elaborate. Sight, sound, smell, pressure, and pain are naturally important, but so are the messages relaying information about the angles and positions of joints, the length and tension of muscles, even the speed of movements.

MOTOR SYSTEMS

The brain's central stalk cradles the brainstem, a control center for breathing, heartbeat, and blood pressure. The area called the medulla, at the brainstem's base, is actually a bulbous extension of the spinal cord, and is a crossroads for one of the two major transmission networks linking brain and skeletal muscles. Beds of nerve fibers swell into twin pyramids on the surface of the medulla, then descend to the spinal cord in nerve tracts known as the pyramidal system.

The pyramidal system is an important mediator of delicate, skilled muscle movements. Its nerve fibers have their cell bodies in the cortex. About 60 percent come from the motor areas and 40 percent from the somatic sensory region. Pyramidal nerves are among the body's

longest, some stretching 2 ft. (0.6 m).

At every step along the descent from brain to muscle, impulses can influence numerous interneurons, and so vary the precision of muscular control. And an average motor neuron may have as many as 15,000 synapses, connections which provide information from all over the body. Parts of the body such as the back, which have a limited precision of motion, are supplied by relatively few pyramidal neurons—perhaps 50,000. Hard muscles, which perform delicate movements, are driven by impulses from roughly 200,000 pyramidal neurons.

The second major transmission network is the extrapyramidal system. Because it produces contractions of groups of muscles either simultaneously or in sequence, it is responsible for larger, more automatic, body movements—such as those of running, walking, and swimming.

TYPES OF MUSCLE TISSUE
There are three different types of muscle tissue: skeletal muscle, smooth muscle, and cardiac muscle. Cardiac muscle is found only in the heart. Skeletal muscle is the most common and accounts for 60 percent of the body's mass. Smooth muscle is found in the digestive system, reproductive system, major blood vessels, skin, and some internal organs.

Muscular System

THE ROLE OF THE CEREBELLUM

No matter how well tuned the muscles, they cannot work correctly without instructions from the brain. This is the role of the cerebellum, the major part of the hindbrain.

BELOW The cerebellum is the major part of the hindbrain and is connected by motor pathways to the cerebral regions above it. Together with the basal ganglia it initiates all movements and maintains muscle tone, the continuous state of slight muscle tension which, during waking hours, keeps all the body's muscles ready for action.

THE MOTOR UNIT

The basic building block for all voluntary movement, which is under conscious control, is the motor unit: a single motor neuron (nerve cell) and all of the muscle fibers it supplies. Each muscle contains many motor units, but the type of muscle determines the size of the units.

Muscles that require precise control and that act rapidly have small motor units of a few muscle fibers, enabling

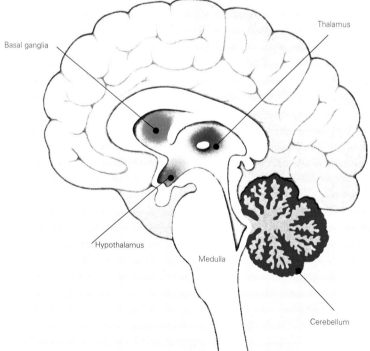

Basal ganglia

Thalamus

Hypothalamus

Medulla

Cerebellum

Muscular System

them to increase tension gradually. In large motor units, tension increases swiftly as the help of each additional motor unit is enlisted. On average there are about 150 muscle fibers to a unit. A single motor unit in the tiny muscles of the eye may contain only three muscle fibers, whereas those in the relatively slow-moving large muscles such as the calf muscles house a thousand or more.

Within each motor unit, muscle fibers obey the "all or none" principle, meaning that all contract or none contract. If the muscle fibers of a motor unit are sufficiently stimulated by nerve impulses to contract at all, they contract maximally.

THE CAUSE OF MUSCLE WASTE

Time or lack of use causes muscle to waste or atrophy; muscular strength usually peaks at the age of 30. As people get older muscle cells degenerate and the number and size of muscle fibers dwindle. Connective or "filler" tissue replaces the lost fibers, making muscles more rigid and slowing their reactions.

DISORDERS AND DISEASES OF THE MUSCULAR SYSTEM

These range from simple problems such as a pulled muscle to muscle wasting diseases with very serious complications. One of these diseases is an inherited disorder, muscular dystrophy. While painless in itself, it is characterized by the

GREATER IN A MAN
Muscles enlarge when you exercise, especially lifting weights, because this stimulates the extra production of actin and myosin which expands the muscle fiber. This is far greater in men than women because it is partly regulated by testosterone, the male sex hormone.

gradual deterioration of muscle function. There are different types of this disorder which affect different areas of the body.

Muscular dystrophy is more common in men than women. Signs of the disease include bulging muscles, which indicate that protein has been replaced with fat cells in the muscle.

With life expectancy reduced, there is no known cure for muscular dystrophy although physical therapy can help to maintain a small amount of muscle function.

Rupture of the Achilles tendon is a common problem encountered by athletes and middle-aged men. Another problem of the Achilles tendon is Achilles tendinitis, which can be caused by over-use of the tendon. Both these problems, while painful, can be cured with painkillers, anti-inflammatory drugs, and rest.

Double vision, resulting in lack of coordinated eye movement is caused when the muscle of the eye is weakened or unable to move due to nerve damage.

CIRCULATORY SYSTEM

Blood is the stuff of life. Like most liquids, it takes on the form of its container—in our case, the human body. It dwells nowhere, has no "home base"—although it belongs within a fraction of an inch of every cell in the body. Blood is red—or, at least, reddish, though at times it looks pink, scarlet, ruddy, or leaden reddish-blue.

It has a characteristic chemical composition—but this varies from minute to minute, and from organ to organ, depending on what its owner is doing.

The appearance of blood signifies life itself—yet the sight of only a few drops outside its container can make grown people faint on the spot. Blood is never still, but always flowing. Nevertheless it goes nowhere, simply round and round in an endless loop. When it ceases to flow, or leaks from the body, so life itself ebbs away.

A RIVER IN REVERSE

The heart pumps blood, and the blood vessels channel and deliver it. Arteries convey blood away from the heart. A surgeon has to be as familiar with the route of each one, and its scientific name, as he or she is with the route to the hospital each morning. Although blood is often likened to a river, carrying its cargo of nutrients to all parts of the body, the arteries are more like a river in reverse. For whereas river tributaries coalesce, and become larger, arteries divide, and become smaller.

Arteries are thick-walled tubes.

A circular sandwich of yellow elastic fibers contains a filling of muscle. The elastic design helps to absorb the tremendous pressure wave of each heartbeat, so that by the time blood reaches the tiny, fragile capillaries it is oozing rather than spurting. The pressure wave is evident after its journey along your arm to your wrist—it is your pulse.

The body's nervous system keeps control of the arterial muscles. By instructing the muscle to contract, the artery's bore is narrowed, and so less blood can flow through. Arteries around the body are continually adjusted, becoming narrower (constricted) or wider (dilated), as the body undertakes various tasks. Run for a bus and the leg muscles need extra blood—so the arteries leading to them become wider. Eat a meal and the intestinal blood supply has to be increased, to absorb digested food—so that it opens up its arteries. Cold? Shut down the vessels near the skin, so that less blood flows there, and less heat is lost. You turn pale, too.

WHAT DO VEINS CARRY?
The veins carry deoxygenated blood back to the heart for revitalization and recycling.

Circulatory System

THIN AND SLACK

Eventually arteries split into arterioles, and then into capillaries, the smallest of the blood vessels. One arteriole may serve a hundred capillaries. In them, in every part of every tissue of every organ, blood's great work is done as it gives up what the cells want, and takes away what they do not want. There is little hindrance: the capillary walls are only one cell wide, and very thin cells at that, looking like curved shingles.

Now the river analogy really does apply. Capillaries join to form venules, to form small veins, to form the main veins (venae cavae), and so back to the heart.

Veins are not at all like arteries. Their walls are thin and slack, because by the time it reaches them blood has almost lost the great pressure which forced it out of the heart. Dark reddy-blue, the blood

oozes slowly on its way. At any one time the veins contain about 75 percent of the body's blood. Some 20 percent is in the arteries, only 5 percent is in the capillaries.

The heart is at the center of the system, filling with blood and then expelling it about once each second, throughout life. Perhaps surprisingly, although heart muscle surrounds so much blood, it has to have its own separate supply. The blood within its chambers is under too great a pressure, and anyway, on the right side of the heart it carries

PLATELETS

A tiny fraction of blood consists of platelets. These are not whole cells but little parts of cells and these help the blood to clot.

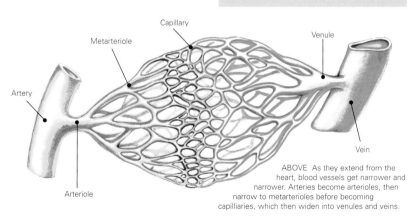

Capillary

Metarteriole

Venule

Artery

Vein

Arteriole

ABOVE As they extend from the heart, blood vessels get narrower and narrower. Arteries become arterioles, then narrow to metarterioles before becoming capilliaries, which then widen into venules and veins.

Circulatory System

little oxygen, being on its way to the lungs for refreshment. So there are special vessels, the coronary arteries, that run across the heart's surface and then branch and dive into its thick, muscular wall. Coronary veins return the favor, by returning this blood to the chambers of the heart. This circuit, quickly around its own pump, is blood's shortest in the body's circulatory system. From the right side of the heart it flows along pulmonary arteries to the lungs, to collect vital oxygen. Then back to the heart's left side, and on around the body once more. The two-part circuit is complete, a sort of figure-eight with the heart at the cross-over.

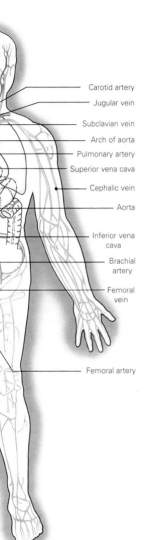

- Carotid artery
- Jugular vein
- Subclavian vein
- Arch of aorta
- Pulmonary artery
- Superior vena cava
- Cephalic vein
- Aorta
- Inferior vena cava
- Brachial artery
- Femoral vein
- Femoral artery

RIGHT A thousand miles of blood vessels keep a constant flow of blood coursing through the human body. This circulatory system enables the heart to pump oxygenated blood along the arteries (red) to the tissues, and to collect waste products in the blood flowing back to the heart along the veins (blue). A shorter circuit carries blood to and from the lungs, to rid it of carbon dioxide and to recharge it with oxygen.

RED CELLS
Red cells are the most numerous cells in the body. In a tiny drop of blood, there are five million of them.

Circulatory System

Absorbing the Breath of Life

Life relies on a supply of energy. In our type of cellular chemistry, liberating this energy from the food we eat is an aerobic process—that is, it requires oxygen. Growth and repair of body tissues depend on nutrients for raw materials, also supplied in our food. The respiratory and digestive systems take care of obtaining oxygen, energy, and raw materials from the outside world. But what about the inside world? Straightaway, blood is on the scene, doing the jobs it does most—fetching and carrying.

In every one of the lungs' 300 million microscopic sacs, or alveoli, blood is only four hundred-thousandths of an inch away from air (give or take a little). It courses through millions of capillaries, that clasp each alveolus like thin fingers. This blood is hungry for oxygen. It has been around, and the body's needs have almost drained it of the vital substance. In the alveoli, oxygen is plentiful, breathed in from the surrounding air and passing down a concentration gradient into every nook and cranny of the lungs.

The efficiency of oxygen exchange is staggering. Spread out flat, the double-layer of alveoli and capillaries would carpet a room 30 ft. (9 m) by 25 ft. (7.5 m). But it would be a thin carpet —the capillaries hold only three ounces of blood at any time. This blood flows sufficiently near to the air for oxygen exchange to take place for only one

quarter of a second.

During this time, oxygen obeys a simple physical law. It moves from a region of higher concentration (the oxygen-rich air) to one of lower concentration (the oxygen-starved blood). It dissolves in the fluid layer lining each alveolus, diffuses through the thin cells making up the alveolar wall and capillary wall, and reaches the blood. Refreshed, and its color turned from leaden reddish-blue to crimson, the blood flows on its way, via the heart, back to the body tissues.

But this is only half the story. As oxygen is required, so carbon dioxide is not. This waste product of bodily chemical reactions has been mopped up from the tissues, in dissolved form, by the blood. It must be jettisoned before its levels creep up and become poisonous. So it is not a one-way street in the lungs, but two-way traffic. Carbon dioxide molecules on the way out trace a reverse path to oxygen molecules coming in. The carbon dioxide, now in gaseous form, is eventually breathed out of the lungs and into the atmosphere.

TWO CIRCULATIONS
There are two circulations in the circulatory system—the systemic circulation and the pulmonary circulation.

Circulatory System

CONSTITUENTS OF BLOOD

When treated with salt, blood settles into three distinct layers, consisting of watery plasma, white blood cells, and red blood cells. The plasma is mostly water containing about 7 percent of dissolved solids in the form of inorganic salts and organic substances. The salts are present as positive and negative ions, while the organics include proteins, fats, and carbohydrates.

RIGHT After being treated with salt the three main constituents of blood are:
watery plasma
red blood cells
white blood cells.

BLOOD TYPES
Different people have blood types called blood groups—there are four main groups, A, B, AB, O. Many people also have so called Rh factor on the red blood cell's surface.

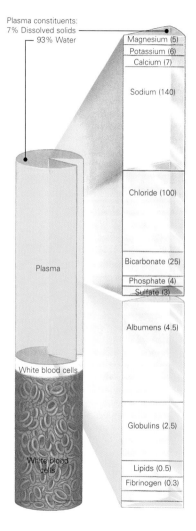

Plasma constituents:
7% Dissolved solids
93% Water

Magnesium (5)
Potassium (6)
Calcium (7)

Sodium (140)

Chloride (100)

Bicarbonate (25)
Phosphate (4)
Sulfate (3)

Albumens (4.5)

Globulins (2.5)

Lipids (0.5)
Fibrinogen (0.3)

Plasma

White blood cells

White blood cells

Circulatory System

THE NUMBERS GAME

Oxygen does not simply float at random, as dissolved molecules washing about in the blood. At least, only around 1 percent of it does. The other 99 percent has a personal carrier: a large protein molecule known as hemoglobin. With a contorted molecular shape resembling a piece of modern, chunky jewelry, hemoglobin consists of some 10,000 atoms that make up four intertwined amino acid chains. Each chain cradles a ring of carbon, hydrogen, and oxygen atoms known as a heme group. Nestling in each heme as a centerpiece is a single atom of iron.

Iron is an "oxygen magnet." It is

BIRTH OF RED CELLS
About three million new red cells are "born," and the same number die, every second.

strong enough to attract oxygen in the plentiful surroundings of the lungs, but not too powerful to keep hold of it in the oxygen-deprived environment of the tissues. On the return trip it returns the favor and transports some of the carbon

BELOW   Blood is the body's principal transportation system, carrying vital materials (pink arrows) to the cells and transporting waste products (yellow arrows) away from them.

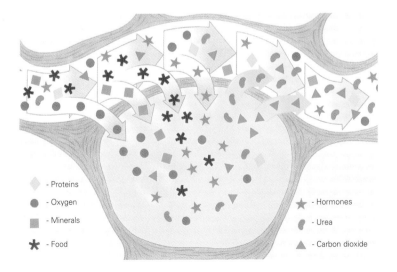

◇ - Proteins
● - Oxygen
▢ - Minerals
✳ - Food

★ - Hormones
◗ - Urea
△ - Carbon dioxide

Circulatory System

dioxide. But most of the carbon dioxide molecules (about 70 percent) enter a chain of chemical reactions that help to maintain the balance of acidic and basic substances in the blood.

In this way each molecule of hemoglobin carries four molecules of oxygen. Again, hemoglobin does not float randomly in the bloodstream. Roughly 270 million molecules of it are jammed together and parceled up in a thin skin to make a scoop-centered, donut-shaped, pinched-disk of a cell. This is the red blood cell, or erythrocyte. It is one of the smallest and simplest cell types in the body, being not much more than a membrane-bounded thick soup of hemoglobin, without a nucleus (control center) and structures found in other cells.

HEMOGLOBIN

Every red blood cell has 300 million hemoglobin molecules, each capable of holding four pairs of oxygen atoms. They pick up the oxygen in the lungs, and surrender it after traveling to the tissues in the blood. A lack of hemoglobin or red cells results in palor and other symptoms of anemia.

RED BLOOD CELL
This is one of the smallest and simplest cell types in the body. With no nucleus, this cell is not much more than a membrane-bound thick soup of hemoglobin.

Oxygen-binding sites (heme groups)

ABOVE A giant hemoglobin molecule. Each molecule consists of four convoluted protein chains, each surrounding an atom of iron.

Circulatory System

SECONDARY CIRCULATIONS — THE HEART AND LUNGS

An important secondary circulation concerns the heart and lungs.

BELOW The oxygenated blood flows out of the lungs to the left side of the heart, while oxygen depleted

blood enters the right ventricle of the heart and is pumped into the lungs to be oxygenated by the alveoli.

ATOMS IN HEMOGLOBIN
Hemoglobin consists of around 10,000 atoms that make up four intertwined amino acid chains.

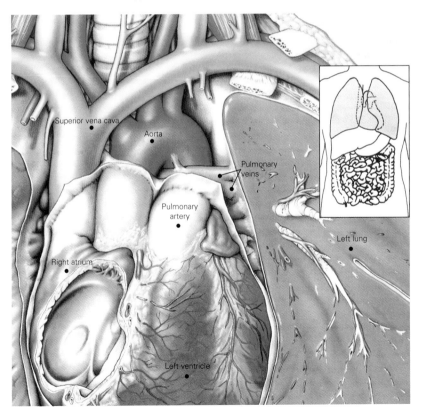

Superior vena cava

Aorta

Pulmonary veins

Pulmonary artery

Right atrium

Left lung

Left ventricle

Circulatory System

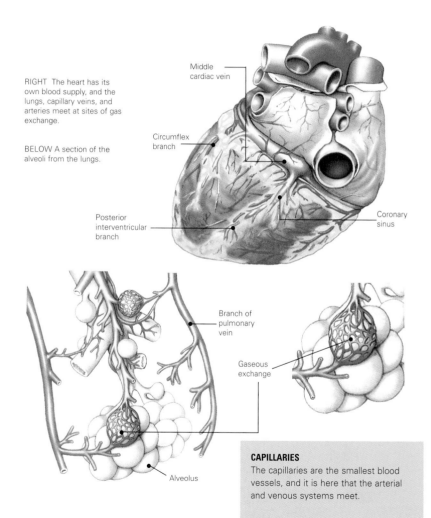

RIGHT The heart has its own blood supply, and the lungs, capillary veins, and arteries meet at sites of gas exchange.

BELOW A section of the alveoli from the lungs.

Middle cardiac vein

Circumflex branch

Posterior interventricular branch

Coronary sinus

Branch of pulmonary vein

Gaseous exchange

Alveolus

CAPILLARIES

The capillaries are the smallest blood vessels, and it is here that the arterial and venous systems meet.

Circulatory System

SECONDARY CIRCULATIONS — DIGESTIVE AND EXCRETORY ORGANS

The other important secondary circulation concerns the digestive and excretory organs which involves the hepatic portal system.

VILLI FROM THE SMALL INTESTINE
Spread out flat, the villi from the lining of the small intestine could cover the size of a tennis court.

BELOW The digestive and excretory organs—the hepatic portal system (shown in purple) carries the products of digestion to the liver.

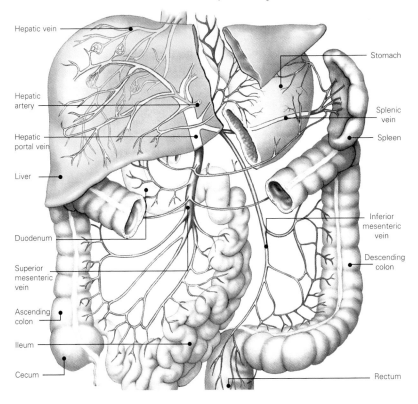

Hepatic vein

Hepatic artery

Hepatic portal vein

Liver

Duodenum

Superior mesenteric vein

Ascending colon

Ileum

Cecum

Stomach

Splenic vein

Spleen

Inferior mesenteric vein

Descending colon

Rectum

Circulatory System

Waste products in the
blood are carried along
the renal artery to the
kidney, which filters
them out along with
excess water to form
urine.

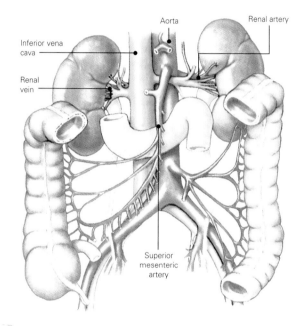

Aorta

Renal artery

Inferior vena
cava

Renal
vein

Superior
mesenteric
artery

RIGHT The excretory organs
showing the renal artery and
kidneys.

BELOW A kidney

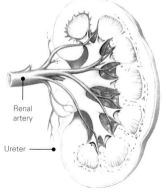

Renal
artery

Ureter

FILTERS IN THE KIDNEYS
There are more than a million delicate
minifilters in each kidney and these are
called nephrons.

Circulatory System

Meal on a Tennis Court

The other half of the energy and raw materials, comes in via the small intestine. Blood on its round-body trip passes into another network of capillaries, this time in the microscopic "fingers," the villi, projecting from the lining of 23 ft. (7 m) of small intestine. Spread out flat, the villi easily outdo the alveoli—they could cover an area the size of a tennis court. Food, broken down by digestion into molecules, along with vital water, passes from the intestinal canal into the bloodstream at the villi, by the process of absorption.

We eat only occasionally, yet our cells need a constant supply of energy and materials. How is this achieved? Blood flows from the capillary network in the villi to a large vessel, the portal vein, that runs from the intestines to the liver. Inside the liver are up to 100,000 tiny hexagonal units, the lobules. They are literally drenched with blood, which is by now low in oxygen, but loaded with newly-absorbed nutrients.

Liver cells in the lobules work their metabolic magic as they convert, store, recycle, and release: glucose, fats, proteins, and other nutrients are dealt with, as and when needed. There is no shortage of supply. Some 30 percent of the blood pumped by the heart in one minute passes through this marvel of a chemical factory. Blood leaving the liver carries the processed nutrients, to be distributed to all tissues.

The liver, unique among organs, has a

> **WHITE BLOOD CELLS**
> White blood cells are of various kinds, each with a particular job to do in the body's defense system. Unlike red blood cells, white cells have a central nucleus.

second blood supply. The hepatic artery brings bright red blood, fresh from the lungs, to deliver much-needed oxygen. This blood eventually mixes with the portal supply, drains into the main hepatic vein, and heads off back to the heart.

Blood Under Analysis

The chemical and particulate cocktail that is blood cannot be dissected by scalpel and forceps, like any ordinary tissue. But in the fourth century BC, the Greek physician Hippocrates observed that a clear flask of salt-treated blood, left to settle of its own accord, separated into three layers. The uppermost is clear and straw-colored; this is plasma. The thin middle layer is made up of white cells. The heaviest layer, red in color and 45 percent by volume, consists of red cells.

Plasma is about 93 percent water—water being the body's solvent. Dissolved in it are any number of substances, from simple salts containing sodium and chloride, to small carbohydrates such as

Circulatory System

glucose, to lipids (fats), to complex proteins such as globulins (many of which are enzymes) and albumens.

At any one moment the vast, pervasive network of some 10 billion capillaries contains 5 percent of the body's blood. Here, in the capillaries, blood gives and takes. It is the great and silent exchange. Oxygen leaves the red blood cells, passes through the single-cell-thick capillary wall, and diffuses through the watery interstitial fluid to reach the cells. It passes carbon dioxide diffusing in the opposite direction. Foodstuffs, too, pass from blood to tissues, while wastes and by-products such as urea do the reverse. Many other body chemicals, such as hormone messengers, join in the great exchange.

WHITE KNIGHTS OF THE BATTLEFIELD

The thin white layer in the flask of settled blood contains an army ready to defend the body to the death—and die they do, in their billions, when an infection or other disease takes hold. The defenders in question are the white blood cells, or leukocytes, outnumbered 600 to 1 by red cells. But, unlike the reds, the whites are "complete" cells, with nuclei and other internal structures. Self-contained and self-sufficient, they are ready to go anywhere at a moment's notice.

There are five major types of leukocytes. When viewed under the microscope, three of them have a grainy,

granular appearance. These are neutrophils (about 60 percent of the total), eosinophils (3 percent), and basophils (7 percent). Their names recall the way each type takes up a standard laboratory stain on a microscope slide. The other two smoother types of leukocyte are lymphocytes (25 percent) and monocytes (5 percent).

BELOW Neutrophil, one of the many types of white blood cell.

RED AND WHITE
White blood cells are also called leukocytes, and are vastly outnumbered by red blood cells. So for every single white blood cell there are 600 red blood cells.

Circulatory System

The leukocyte army is a team of specialists. Half the circulating army patrols in the blood, while the others are out and about in the tissues, checking that all is well. White cells are mobile. They can creep amebalike along capillaries, squeeze out of them through gaps between the cells of their walls, and pour themselves along spaces between the cells of the tissues.

Invaders are the enemy. They may be bacteria, viruses, fungi, or parasites. Somehow they always seem to be getting in, through cuts in the skin or through the delicate linings of the respiratory and digestive tracts. When they appear, the army swings into action. Basophils and some lymphocytes act like mines, "blowing up" and releasing chemicals which trigger the disease-coping inflammatory processes.

Neutrophils, eosinophils or monocytes rush to the battlefield and literally gobble up the invaders, enveloping them in their cellular folds and absorbing them in a procedure termed phagocytosis (from Greek words meaning "cell eating"). The phagocytes crowd into the inflamed area, devouring as much as they can; they also set off through the body tissues to mop up invaders that have escaped the bloodstream.

Leukocytes are not the longest-lived of body cells. Most are in action for only a few hours. In healthy blood their numbers are low. Yet reserves are waiting, in the bone marrow, lymph nodes, and spleen, where they live for a week.

DIFFERENT WHITE CELLS
There are many different kinds of white blood cells each with a particular job to do in the body's defense system, and unlike red blood cells, white cells have a central nucleus.

ABOVE Lymphocyte, a white blood cell.

Circulatory System

At the first sign of trouble, these reserves
march to the battlefield. At the same time
the white cell production line moves up a
gear and hurriedly manufactures required
new troops. It is this heated production
that causes the fever of illness, while
aching bones and a sore throat (from the
swelling of the lymph nodes) shows that
the body is busy fighting back.

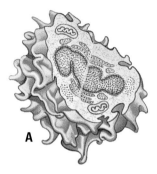

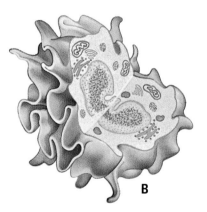

ABOVE, LEFT and BELOW are three other
kinds of white blood cells:
A – Eosinophil
B – Monocyte
C – Basophil

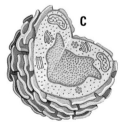

PHAGOCYTES
Among the many kinds of white blood
cells in your blood are ones called
phagocytes. These destroy bacteria by
surrounding and eating them. Pus forms
if the bacteria kill the phagocytes.

Circulatory System

WHEN BLOOD AND AIR MEET

Blood and air meet in the capillaries of the lungs. Incoming blood sheds its load of waste carbon dioxide, while its hemoglobin latches on to oxygen gas and carries it away.

ALVEOLUS
The tiny grapelike structure in the lungs where gas exchange takes place is called the alveolus. The total surface area of all the millions of alveoli is huge.

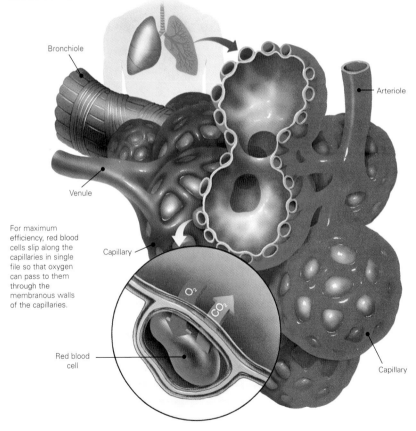

Bronchiole

Arteriole

Venule

For maximum efficiency, red blood cells slip along the capillaries in single file so that oxygen can pass to them through the membranous walls of the capillaries.

Capillary

O_2

CO_2

Red blood cell

Capillary

Circulatory System

THE SECRET ARMY

How do white cells distinguish friend and foe—"self" from "non-self"? Why do they not turn on each other, or attack red cells or some other part of the body? The answer lies in the almost mystical sub-world of immunology. We must descend below tissue and cell level, to the very molecules which make up cells—more specifically, to the molecules which link together to form the "skin," be it of a bacteria, virus, leukocyte, or any other speck of organic matter.

The white army recognizes its body's own cells. Their molecular coats are of known color and pattern, and signify allegiance. But the microbial invaders wear coats of different colors and patterns. The units of which these unfamiliar coats are constructed are known as antigenic determinants, and their wearers are called antigens. To the white army, antigens are the signal to attack.

The various types of phagocytes follow the signal and simply consume invaders, of any type, alive or dead. But the lymph-ocytes are altogether more cunning. They are the secret army within an army, the stealth force. There are two main divisions, T lymphocytes and B lymphocytes. Under a microscope they look alike, yet their roles are different, although complementary. They do not all enter the fray. These cells are individually pre-programed to recognize a specific "fingerprint" of antigenic substances

> **PRODUCING ANTIBODIES**
> The white blood cells, lymphocytes, in the fluid produce antibodies that attack diseased organisms and "switch on" the immune system to fight foreign invaders.

carried by one particular invader, from a common cold virus to the bacterium of tuberculosis.

T lymphocytes themselves come in three varieties. Some are killers; they attack the invaders directly with potent chemicals. Others are suppressors; they help to regulate the fight back, protecting the body from the excesses of its own defense. The third type are helpers; they prod the B lymphocytes into action. (The "T" stands for thymus, the gland behind the breastbone that processes these lymphocytes.)

THE MAGIC BULLETS

B lymphocytes make antibodies. These are not another cell type. Antibodies are the body's "magic bullets," each with the identity of a specific antigen engraved on it. They are protein molecules craftily shaped to stick onto, disrupt, and disable intruders. They form the cornerstone of the immune response.

The scenario for war might run as follows. Suppose a virus dares to enter the body. The lymphocytes programed to detect it are activated. They might number only a few dozen, in the millions

Circulatory System

of lymphocytes hanging around in a lymph node. When they meet their antigens, they stick or "bind" to them in mutual recognition. Immediately, the B cells are transformed. Many go on to clone from themselves the plasma cells, which are antibody factories. Within a few days, one lymphocyte has multiplied

THE BODY CAN BE TRICKED
By giving the body a weakened or disabled potion of microbial invaders, the immune system reacts as though the fake invasion was real. We become immune, or resistant, without suffering from the infection. This "potion" is a vaccine that has been used to save millions of lives around the world.

BELOW Antibodies are Y-shaped protein molecules that stick onto invading bacteria or form bridges between them so that they clump together. This clumping, or agglutination, makes bacteria easy targets or phagocytes.

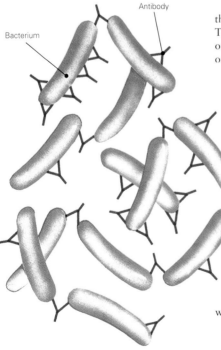

Antibody

Bacterium

into hundreds of plasma cells, each of which mass-produces antibodies at the rate of around 2,000 every second.

The antibody is shaped like a Y, with the two arm tips as the business end. These lock on to antigenic determinants on the invaders—destroying them, and only them.

Agglutination helps. This is "clumping" of the antigens, facilitated by antibodies. Each arm of the Y antibody may attach to a different antigen, so linking the two together. Antibodies are generally sticky molecules, in any case, and sometimes glue themselves to other proteins or tissues. This process quickly snowballs. In no time, antigens too small to be noticed—some even small enough to "hide" by staying in solution—become knotted together and coated by antibodies. They are now bigger, easier targets for those hungry scavengers, the macrophages and neutrophils among the white cells.

Circulatory System

SOLIDS OUT OF THE LIQUID

Blood functions in many ways once invaders are in the body. It is also in action as they try to gain access. Blood clots, or coagulates, to seal a leak, to save itself from losing itself, to retain vital body chemicals, and to help repel the ever-waiting invasion of body hijackers from without.

When a blood vessel is bruised, cut, or otherwise injured, three mutually reinforcing processes move into action. The damaged vessel contracts, to restrict the flow of blood and therefore minimize any loss. Second is formation of a platelet plug. Platelets are "almost cells"—they resemble cell fragments, rather than whole cells. Only one-quarter the size of red cells, they float in the blood plasma, ever ready to seal a break. They live for about ten days, and an average adult makes 200 billion new platelets every day.

FAMILIES OF BLOOD

Agglutination, and a sort of antigen, figure in another feature of blood, which has also saved countless lives. People have long had the notion of giving some of a healthy person's blood to one with greater need—a badly-bleeding accident victim, for example. Early on, such transfusions were hit-and-miss affairs. Sometimes they saved, but sometimes they worsened the situation, because the red cells inexplicably and unpredictably clumped together and clogged up the recipient's circulation.

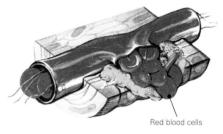

Red blood cells

ABOVE An injured blood vessel

At the beginning of the twentieth century a young Viennese researcher, Karl Landsteiner, set out to prove that there were individual differences in human blood. He succeeded, and in 1930 was awarded a Nobel Prize.

Landsteiner and his colleagues discovered blood groups, or blood types. They took samples of blood from different people (including themselves) and separated them into plasma and red cells. Then they mixed various red cells with different plasmas, and noted what happened: in some mixtures the red cells clumped together, in others they did not.

From this simple beginning, the intricacies of the ABO system of blood types were unraveled. Red cells, like other cells, carry a specific molecular pattern on their surfaces. The pattern includes a type of molecule called an agglutinogen—an antigen that triggers agglutination, or clumping. There are two types of agglutinogen—A and B. Some people have only A; others have only B; some have both (AB); and some have none (O).

Circulatory System

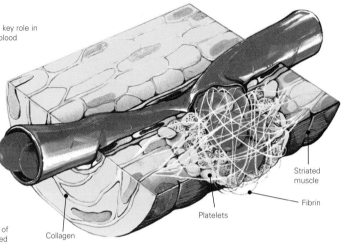

RIGHT Platelets play a key role in stemming the flow of blood from an injured vessel. As soon as a blood vessel hemorrhages, platelets cling to collagen in the vessel wall and release adenosine diphosphate (ADP). This substance acts as a signal that calls other platelets to the wound. Another chemical from the damaged tissue, thromboplastin, converts the protein fibrinogen into strands of fibrin, which enmesh red cells and bind them into a plug.

Striated muscle

Fibrin

Platelets

Collagen

The plasma reciprocates. People with A agglutinogens in their red cells (that is, A type blood) have anti-B agglutinin, a type of protein, in their plasma. If red cells carrying B agglutinogens ever get into A type blood, the anti-B agglutinin in the A plasma reacts quickly with them and prompts them to clump, gluing them together like piles of sticky donuts.

People with B agglutinogens on their red cells have anti-A agglutinin in their plasma. Those with type O blood (neither A nor B agglutinogens) have both anti-A and anti-B agglutinins. O is the "universal donor" because, lacking agglutinogens, it cannot trigger clumping no matter what is in the plasma. Similarly, AB is the "universal recipient."

Landsteiner's original work has been much expanded. Today physicians recognize ABO, Rhesus, and dozens of other systems of typing blood. It may be that a person's "bloodprint" is as unique as his or her fingerprint.

NATURAL BAND AID
Although it looks firm, a newish blood clot is 99% water. Within minutes it starts to contract and squeezes out a pale fluid called serum. When exposed to the air, this clot hardens to a brittle bump—the body's natural band aid.

Circulatory System

BLOOD TYPES

The world distribution of blood types provides anthropologists with clues about migration of ancient mankind. The map below shows the present-day distribution of type B, which is commonest in central China. Its incidence spreads westward through Europe, possibly reflecting the thirteenth-century Mongol invasions, and eastward across the Bering Strait into North America, indicating a common ancestry between American Indians and Mongolian peoples. The compatibility of blood types is of crucial importance in blood transfusions.

RIGHT As the compatibility of blood types is of crucial importance in blood transfusions, the types that match and can safely be mixed are shown in pink in the diagram. Donor types (D) are listed in the left-hand column, with recipient types (R) along the top of the chart.

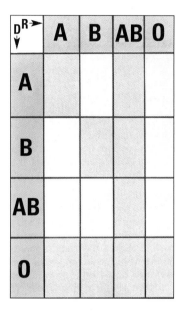

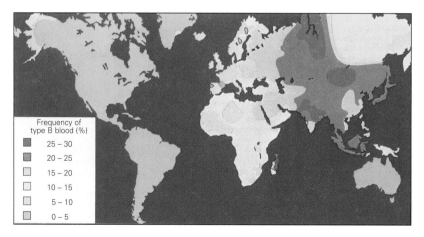

Frequency of type B blood (%)

- 25 – 30
- 20 – 25
- 15 – 20
- 10 – 15
- 5 – 10
- 0 – 5

Circulatory System

Any blood type can be mixed with the same type (A with A, B with B, and so on), either A or B can be mixed with type AB, and type O—often called the universal donor—can safely be mixed with any other type. But an attempted transfusion in which type B is donated to a recipient of blood type A, for example, would result in clumping and dangerous consequences for the recipient.

LYMPH: BLOOD'S POOR RELATION

Blood is not the only liquid tissue. Its "mirror-image," perhaps more a poor relation, is lymph. Lymph derives from blood, and returns to it. The fluid and white cells that leak from the capillaries must go somewhere. They drain from a nebulous network of ill-defined spaces into a more organized system of channels, which join together in a reflection of the capillary and venous sections of blood's circulatory system. The channels are lymphatic vessels, and the fluid they convey is lymph. It contains white cells galore, and also proteins, fats, and other products of digestion. The villi that line the small intestine are rich in lymph vessels called lacteals, and these are especially important in the absorption of fats.

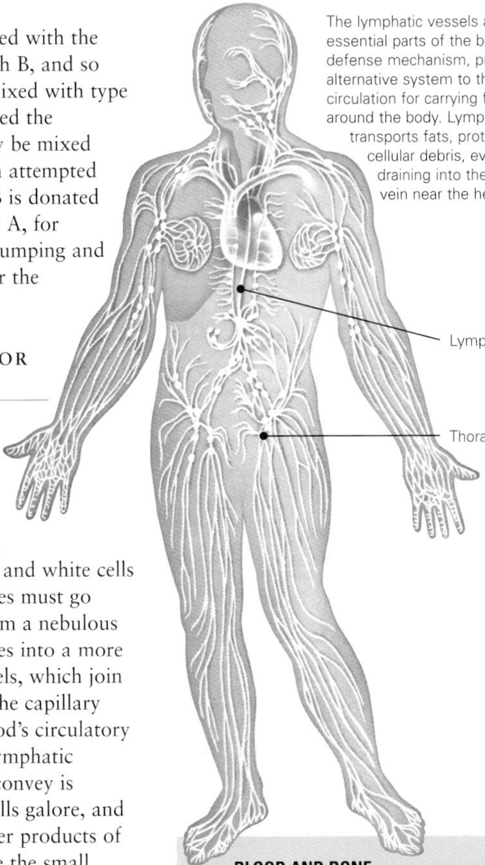

The lymphatic vessels and ducts, essential parts of the body's defense mechanism, provide an alternative system to the blood circulation for carrying fluid around the body. Lymph also transports fats, proteins, and cellular debris, eventually draining into the thoracic vein near the heart.

Lymph node

Thoracic duct

BLOOD AND BONE
In an adult, the marrow of the skull, ribs, and spine makes most red cells, but any bone marrow in the body can be called upon to contribute in an emergency.

RESPIRATORY SYSTEM

Speech, from a whisper to a shout, and the subtle sense of smell, also depend on the movement of air in the respiratory passages. Respiration, with digestion, is one of the body's two great "input" processes.

The respiratory system's main job is to absorb oxygen from the atmosphere into the body, and to expel the waste product carbon dioxide from the body into the atmosphere. Its various parts take up much of the space in the face and the neck, and most of the chest. The system occupies a lot of physical space and expends hard-won energy in the muscle-powered movements of breathing.
Over thousands of years of evolution, our respiratory system has become adapted to perform various other functions. Breathing carries odor molecules on the incoming airstream, which are detected by the olfactory organs in the nose and perceived by the brain. Perhaps the odor warns us off a potential meal, signaling that food is bad and not fit to eat, or advertises the presence of a mating partner. The respiratory airflow has also been "hijacked" by the larynx (voice-

box), which exploits it to create the vast range of sounds that makes humans so communicatively vocal.

VITAL OXYGEN

The respiratory system's role is fundamental to life. The oxygen it obtains is supplied via the circulatory system—the heart and blood vessels—to every cell in the body. Without oxygen, cells die. Within minutes of oxygen deprivation, tissues become irreparably damaged. This is because oxygen is a vital ingredient of metabolism, the biochemical processes that keep the body running. Digested nutrients, from food, are mixed with oxygen, enzymes, and other chemicals. After several intermediate stages of chemical reaction, the result is a helping of available energy. The cells "burn" this energy, using it to drive its many other biochemical processes. It is, literally, "energy for life."
   The biochemical process of burning energy, using oxygen in this case, is often called aerobic or cellular respiration. The physical process of taking in air in order to absorb the oxygen in the first place is also referred to as respiration. They are the end and beginning, respectively, of the same overall process.
   The heart is often regarded as being "up front," the foremost organ. We all

BRANCHES LIKE TREES
Like a pair of trees, the airways and blood vessels divide and branch within the structures of the lungs. The windpipe, splits into two main bronchi, one leading to each lung. These bronchi divide and subdivide to form the bronchioles.

Respiratory System

know that if the heart stops, life ceases. Yet the pumping action of the heart simply distributes to all corners of the body the oxygen already obtained by the respiratory system. The pulsing blood collects the carbon dioxide that the respiratory system ultimately expunges from the body. No matter how hard and fast the heart pumps, if there is no fresh oxygen, there is soon no life. Even the very muscle of the heart itself relies on this continuing supply of oxygen. Small wonder we talk of "the breath of life."

NOT IDENTICAL
The two lungs are not quite the same size as each other. The right lung is larger than the left lung and is sectioned into three lobes, whereas there are only two in the left.

TREASURED CHEST

While you have been reading, you have certainly been doing at least one other thing; breathing. Air has passed in and out, probably at the rate of about 14 breaths a minute, unless you have just completed a run around the block or been watching a torrid love scene on the television. Physical effort and emotional passion both bring on heaving chests, for within the chest reside the two lungs, the twin centers of the respiratory system.

The system can be divided into three main sections. The first contains the "pumping machine," muscles and elastic fibers which pull and push air in and out of the lungs. The second is the system of conducting airways through which air is transported to and from the third region, the lung alveoli. These microscopic "air bubbles" within the lung tissue facilitate absorption of the vital oxygen into the bloodstream.

RIGHT The trachea or windpipe splits into two main bronchi. Each of these two bronchi lead to the lungs where they divide and subdivide to form bronchioles—the narrowest airways that end in tiny alveoli.

Respiratory System

The lungs are large, cone-shaped, pink-gray, spongy organs. The two lungs are not quite mirror images of each other. The right one is larger and is sectioned into three main compartments, or lobes. The left lung has two lobes and a hollow scooped in its lower front, in which nestles the ever-pounding heart. Indeed, the lungs themselves are never still. In quiet, restful breathing, they draw in (and push out!) about three-quarters of a pint of air with each breath. That's around 15 pints each minute, or 790 gallons overnight—about one-third of the air in a smallish bedroom. In the gasping aftermath of a 1,000-meter race, the lungs' throughput is 20 times their resting turnover.

Air is conveyed down into the lungs by a branching system of tubes that have been likened to an upside-down, hollowed-out tree. The base of the bronchial tree consists of the nose and mouth, twin inlets for atmospheric gases. The main "trunk" is the throat and the trachea, or windpipe, a 5-inch-wide tube which links the throat to the chest. The trachea divides into two smaller tubes, the principal bronchi, one to each lung. These tubes further subdivide into smaller and smaller bronchi (the "branches" of the tree). Ultimately the smallest air passages, the bronchioles (the "twigs"), end in the minute air sacs called alveoli—more than 350 million of them in each lung. Surrounding and supporting the airways and alveoli are elastic fibers and connective tissue, the interstitial tissue or ground substance of the lungs. The whole gives the lungs a somewhat spongy texture.

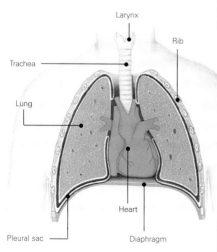

Larynx

Rib

Trachea

Lung

Heart

Pleural sac

Diaphragm

LEFT In this cutaway view of the chest cavity, the upper airways can be seen leading down to the lungs behind the heart. The lungs are surrounded by the double membrane of the pleural sac, and supported by the muscular dome of the diaphragm.

THE COLOR RED
The blood contains different cells—red and white cells and platelets. It is the red cells which contain a substance called hemoglobin—the body's oxygen carrier—which colors the red cells in blood.

Respiratory System

THE PISTON OF THE RESPIRATORY PUMP

The stretchy, spongy structure of the lungs permits them to continually enlarge and contract, to suck in air and blow it back out; that is, to breathe. The pumping system that powers the lungs involves the bony cage around the chest created by the spine, ribs, and breastbone (sternum), and two sets of muscles.

The first set is the diaphragm, a muscular sheet that consists of three overlapping groups of crisscrossed muscle fibers. The diaphragm forms the floor of

EACH LUNG
Being the smallest air passages, the bronchioles end in the minute air sacs called alveoli. There are more than 350 million of these alveoli in each lung.

the chest, or thorax, and separates it from the abdomen below. Its edges connect to the spine at the back, to the lower ribs around the sides, and to the bottom of the breastbone at the front.

A relaxed diaphragm is domed, almost bell-shaped. Its center bulges under the pressure from the abdominal organs below and projects up into the chest, to a level only an inch or so below the nipples.

A tensed, contracted diaphragm is much flatter and its net result is to enlarge the chest, pulling down on the lungs and increasing their volume. It works rather like a piston moving down inside a cylinder.

The lungs inflate because each is wrapped in a thin, slippery membrane, the pleura, which folds back on itself to line the inside of the chest. The part covering the lung is known as the visceral pleura; that lining the inside of the chest is the parietal pleura. The two layers of the pleura are in intimate contact, lubricated by an incredibly thin smear of oil-like pleural fluid between them. The entire set-up is an airtight fit in the chest cavity. The only way in or out is along the trachea.

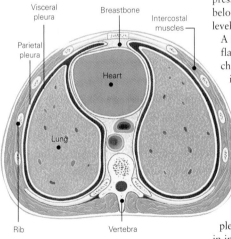

Visceral pleura — Breastbone — Intercostal muscles — Parietal pleura — Heart — Lung — Rib — Vertebra

ABOVE This cross-section of the chest reveals how the lungs are protected within the bony cage of the ribs, spine, and breastbone. See how the left lung is slightly smaller than the right, to allow room for the heart.

Respiratory System

FLEXIBLE CAGE

The second set of breathing muscles consists of those in the back, neck, and, especially, the intercostal muscles between the ribs. As they contract in coordinated unison, they pull the ribs upward. Joints, between the ribs and spine at the rear and the ribs and breastbone at the front, allow the cage to swing up and also out, due to the slightly down-slanting position of the ribs at rest. Once again, the net result is to enlarge the chest, pulling the lungs forward and increasing their volume.

Breathing in—inhalation, or inspiration—is therefore an active process. It requires muscle power from the diaphragm, the intercostals, and other chest-wall muscles. The lungs themselves play a purely passive role in the inspiratory movement: the changes that they undergo are brought about solely through changes in the capacity of the chest. Almost two-thirds of the volume of air inhaled in an average breath is brought about by the pistonlike action of the diaphragm; the rest comes from the motion of the rib cage.

Breathing out—exhalation, or expiration—is a simpler process. The diaphragm and chest muscles relax. The chest cavity elastically recoils to its previous volume. It is aided by the elasticity of the lungs, which expels air from them, and abetted slightly by gravity pulling down on the ribs. As the lungs contract they blow air back up through the trachea.

REMEMBERING TO BREATHE

The movements of breathing are automatic. We can impose our will on them and change them when we want to—for example, during speech or when coughing. But for most of the time, breathing is controlled in a reflex manner by the respiratory center in the brain.

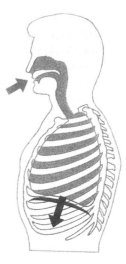

IN ONE MINUTE
The average person breathes in about ten to fourteen times in one minute.

Respiratory System

The respiratory center is situated in the brainstem, or medulla oblongata, a "primitive" region of the brain concerned with vital processes such as heartbeat, breathing, and blood pressure. The respiratory center ensures, no matter how occupied the brain is elsewhere—whether concentrating on a symphony, full of rage at injustice, or simply asleep—that we always remember to breathe. It happens with hardly a moment's thought, every few seconds, every day, throughout life.

Actually, the above is not quite true. The first nine months of life, in the womb, are an aquatic existence. The fetus floats in a pool of fluid, which filters into the lungs. There is no air to breathe. But this creates no problem: oxygen, nutrients, and other essentials are passed to the developing fetus via the placenta, which is both its lungs and its intestines. This situation requires a specific blood flow around the body, which must change after birth when the lungs become functional.

AIR
Breathed-in air contains almost 21 percent oxygen, while breathed-out air contains about 17 percent oxygen.

BREATHING IN AND OUT

Breathing in and out involves slight movement of the ribs and larger movement of the diaphragm.

LEFT As the diaphragm muscle contracts and moves downward, the lungs enlarge and air rushes into them through the respiratory tract.

RIGHT Upward movement of the diaphragm forces air out again.

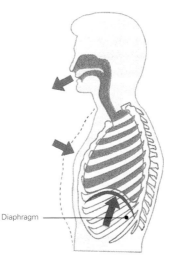

Diaphragm

Respiratory System

BREATHING VIA THE PLACENTA

The heart is effectively two pumps side by side. The right pump sends blood to the lungs for oxygen; this "refreshed" blood returns to the heart's left pump, from where it is sent around the body to deliver its oxygen to the tissues. The blood then returns to the right pump again, so completing the double circuit.

Fetal blood circulation is different. During the first four or five months of development, the fetus has a "hole in the heart," between the upper chambers (atria) of the left and right pumps. This allows most blood returning to the right pump, from the body, to flow directly through into the left pump and so bypass the lungs. The lungs receive only a small "maintenance" supply to fulfill their basic requirements, much like any other organ.

As the fetus grows, the hole gradually closes up, and normally it has gone by birth. This forces more and more blood into the pulmonary artery, the main vessel that conveys it from the heart's right pump to the lungs. However, at this stage there is another bypass, the ductus arteriosus. This small valvelike tube links the pulmonary artery to the aorta, the main vessel carrying blood from the left pump to the body. Thus most blood leaving the heart's right pump, apparently destined for the lungs, is shunted into the aorta and off around the body again; it is another kind of pulmonary bypass.

CIRCULATION OF THE FETUS

While the fetus is in the womb, blood from the placenta flows first to the right side of the heart and then, via the oval foramen, to the left side and on around the body.

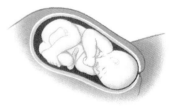

ABOVE A fetus in the womb receives oxygen in the blood from the placenta.

HOLE IN THE HEART
For the first few months of development in the womb, the fetus has a "hole in the heart," but as the fetus grows, this hole gradually closes up and is gone by the time the baby is born.

Respiratory System

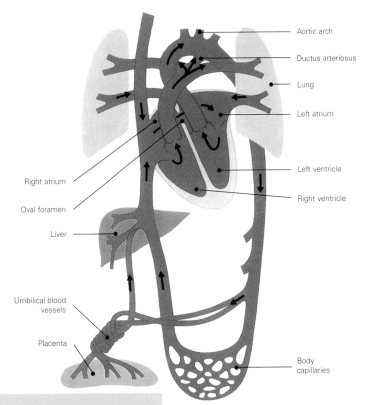

Aortic arch

Ductus arteriosus

Lung

Left atrium

Right atrium

Oval foramen

Liver

Left ventricle

Right ventricle

Umbilical blood vessels

Placenta

Body capillaries

BREATHING BEFORE BIRTH
The fetus is known to make small breathing movements during the last weeks in the womb. One effect that these practice breaths have, is to squeeze excess fluid out of the lungs in preparation for the outside world.

ABOVE While it is still a fetus in the womb, blood from the placenta flows first to the right side of the heart and then, via the oval foramen, to the left side and on around the body.

Respiratory System

THE FIRST BREATH

Before birth, the fetal lungs are not in a collapsed state. They are "inflated" to about two-fifths of their total capacity by a special fluid produced by alveolar cells. The fluid contains a surfactant, a form of natural detergent. It is thought that surfactant helps the baby to overcome the tremendous forces of surface tension in the alveoli, where air must replace liquid in the first breaths. The fetus is known to make small breathing movements in the few weeks before birth, and one effect of these practice breaths is to squeeze excess fluid out of the lungs.

At birth, the baby's diaphragm is jolted into action and the first breath is drawn—usually accompanied by loud and lusty crying. It involves an enormous effort on the part of the new infant. As air is pulled down into the lungs, surfactant helps to keep the alveoli open, and excess lung fluid is absorbed into the bloodstream. Surfactant continues to play this life-giving role through life.

Within a few minutes, the ductus arteriosus begins to contract and should close completely between the fourth and tenth day after birth. It is not known for certain what causes the ductus to close, but it probably involves the contraction of smooth muscle tissue in its walls, and also particular hormones called prostaglandins. If all goes well, the ductus gradually shrinks and becomes a fibrous band joining the pulmonary artery to the aorta, which persists throughout life.

Birth also sees the shrinking of the umbilical arteries and vein, which formerly transported blood to and from the placenta (afterbirth) along the umbilical cord. Modern obstetric procedure may hasten the process by tying and cutting the cord as soon as the baby's breathing becomes established.

CIRCULATION OF NEWBORN BABY

A vital change takes place in a baby's blood circulation at the moment of birth.

The baby is no longer an appendage of the mother, but has started out on the road to independence. The first breaths are hard work, and it takes several days before uniform lung ventilation is

---

AFTER EXERCISE
After lots of exercise, such as running a race, the breathing rate may increase to 50–60 times each minute, and the air that is breathed in and out can amount to a gallon or more.

Respiratory System

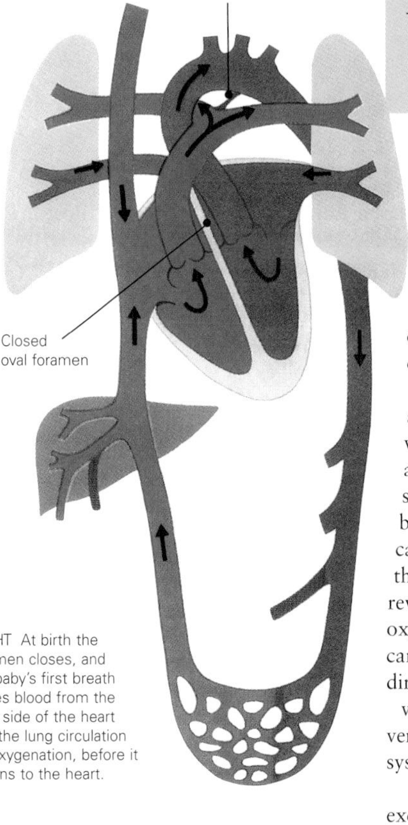

Closed ductus arteriosus
Closed
oval foramen

RIGHT At birth the foramen closes, and the baby's first breath forces blood from the right side of the heart into the lung circulation for oxygenation, before it returns to the heart.

TWIGS
The bronchioles, or the slenderest "twigs on the tree" are less than one twenty-fifth of an inch (millimeter) across.

achieved. What follows are some 500 million or more breaths, until the last one signals the end of life.

The postnatal pattern of blood flow is slightly more complex. Blood from the right ventricle, the main chamber of the heart's right pump, surges out through the pulmonary artery to the lungs. There it takes on board the vital oxygen. Then back it flows via the pulmonary veins to the left atrium, the smaller reception chamber of the left pump. This part of the circuit is termed the pulmonary circulation.

The blood then passes from the left atrium through a valve to the left ventricle, which drives it powerfully along the aorta and its branching system of arteries to the rest of the body. In the thousands of miles of capillaries, the smallest vessels that thread through all the tissues, the rewards of respiration are finally seen: oxygen diffuses from blood to cells, and carbon dioxide passes in the opposite direction. The capillaries unite to make veins, which return the blood along the vena cavae to the right atrium. This is systemic circulation.

The circulatory system permits gas exchange in the capillary network: oxygen is traded for carbon dioxide. The workings of the respiratory system involve its own capillary network, in the lungs, doing the opposite. First, however, air must permeate deep into the lungs, to reach the capillaries.

Respiratory System

Inspired air usually enters the upper respiratory tract (nose, throat, and trachea, or windpipe) through the nostrils rather than the mouth. The nose has advantages as the inlet, for a number of reasons. The nose filters: hairs in the nostrils trap dust and other particles that float in even the cleanest air. In addition, the sticky mucus lining the nasal cavity behind the nose also entraps airborne matter. And the nose warms: the copious blood supply to nostrils and nasal cavity (witnessed by the seeming "flood" of a nosebleed) passes heat to the incoming air. And the nose humidifies: moisture from the mucous lining evaporates and dampens the airstream. These built-in nasal features deliver air more suited to the lung enviroment.

THE SITE OF SPEECH

After it leaves the nasal cavity, the air passes backward and downward through the larynx, or voice-box. This complex three-dimensional structure of cartilage plates, thin bone, muscles, and ligaments is the site of speech. In its center,

projecting from its inner walls, are two pearly, stiffish folds of tissue. Each of these vocal cords, or folds, is about two-thirds of an inch long. Every sound we make, from the softest whisper to an unbridled scream of pain or anger, emanates from the vocal cords.

There is a gap between the cords which allows air to pass silently to and fro during quiet breathing. As we prepare to speak, muscles pivot cartilages that swing the cords together. Air flowing through the slit between them now sets up vibrations along their length.

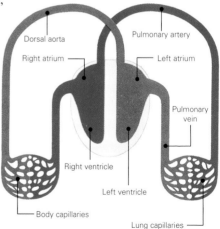

ABOVE The human body has two blood circulations, called the pulmonary circulation (shown on the right of the diagram) and the systemic circulation (on the left). Blood low in oxygen is shown blue, oxygenated blood shown red.

Respiratory System

A TREE OF BRONCHI

Immediately beneath the larynx is the trachea. This gateway to the lungs, the beginning of the lower respiratory tract and the "trunk" of the bronchial tree, is similar to the air hose from a portable hair drier. It is flexible, to cope with bends and twists as the head and neck move. It has strengtheners in its wall, to maintain an open airway even under pressure (in the body, pressure is exerted by muscles in the chest and by the lungs themselves). Unlike the wire helix of a hair drier hose, the tracheal stiffeners are 16 to 20 C-shaped hoops of cartilage.

At its lower end, the trachea forks into two short, stubby tubes, the first of the bronchi. More forks occur as the bronchi, the "branches" of the tree, become narrower. Cartilage stiffeners hold open the larger bronchi, but they gradually become

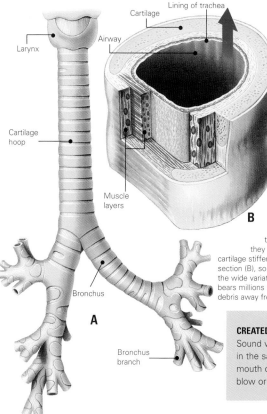

LEFT The major airways of the upper respiratory tract (A) are the trachea, or windpipe, and bronchi. The trachea extends from the larynx to the fork of the main bronchi, which subdivide as they extend into the lungs. Hoops of cartilage stiffen the trachea, as shown in the cross-section (B), so that the airway remains open even with the wide variations of air pressure inside it. The lining bears millions of hairlike cilia, which move mucus and debris away from the lungs.

Labels in figure:
- Larynx
- Cartilage
- Lining of trachea
- Airway
- Cartilage hoop
- Muscle layers
- Bronchus
- Bronchus branch
- A
- B

CREATED BY VIBRATIONS
Sound waves are created by vibrations, in the same way that the reeds of a mouth organ tremble noisily as we blow or suck wind past them.

Respiratory System

more sparse until, some 20 divisions after the trachea, they disappear altogether. These slenderest "twigs" on the tree, less than one twenty-fifth of an inch (a millimeter) across, are the bronchioles. Their structural integrity is derived from muscle fibers that run around their diameter like so many rubber bands.

Some germs, dust, and dirt particles evade the nasal filtering mechanisms. But the pipes and passageways of the lower respiratory tract are fully prepared. All of these airways are lined with mucous membranes. Special cells in the membranes known as mucous or goblet cells, make endless supplies of sticky mucus, which catches and holds settling particles. The membranes also bear countless millions of tiny hairlike cilia. Under the microscope, the cilia can be seen beating upward like teams of rowers wielding miniature oars, driving the tide of mucus and its flotsam of trapped refuse relentlessly up and out of the lungs, to the throat. Every so often we swallow the mucus, perhaps with a small cough to "clear the throat"—or rather, to clear the lungs.

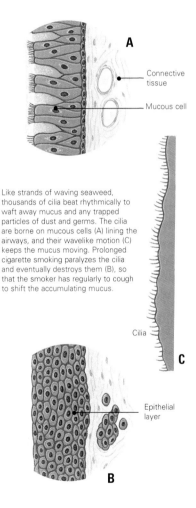

A

Connective tissue

Mucous cell

Like strands of waving seaweed, thousands of cilia beat rhythmically to waft away mucus and any trapped particles of dust and germs. The cilia are borne on mucous cells (A) lining the airways, and their wavelike motion (C) keeps the mucus moving. Prolonged cigarette smoking paralyzes the cilia and eventually destroys them (B), so that the smoker has regularly to cough to shift the accumulating mucus.

Cilia

C

Epithelial layer

B

WHY WE NEED TO BREATHE

All the cells in our body need oxygen, so without it they wouldn't be able to move, build, reproduce, and turn food into energy. Therefore without oxygen they and us would die.

Respiratory System

GRAPES ON THE VINE

If the bronchial tree can be likened to a vine, then the terminal bronchioles are the final twigs, and the alveoli the grapes. Each bronchiole ends in a cluster of these minute air sacs. The name of the game is now surface area, because the greater the area the lungs have for gas exchange, the higher is their oxygen-absorbing

efficiency. In fact, compared to their physical size, the lungs have an immense area of between 50 and 100 square yards. This is nearly the size of a tennis court, neatly folded and compactly bundled into the chest cavity.

The pulmonary artery, bringing deoxygenated (low-oxygen) blood from the heart's right pump, divides into two branches, one for each lung. In yet another version of the branching tree, the artery divides many times until the vessels become capillaries. A network of these pulmonary capillaries surrounds each alveolus. Blood flows steadily through a conveyor belt bringing carbon dioxide for disposal and collecting oxygen for dispersal. At any instant, around one-fifth to one-tenth of the body's total blood volume, ten and a half pints, is in the pulmonary circulation.

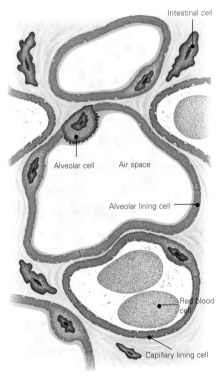

Intestinal cell

Alveolar cell   Air space

Alveolar lining cell

Red blood cell

Capillary lining cell

LEFT For efficient gas exchange to take place in the lungs, the alveolar air spaces, and the circulating red blood cells must be very close together, as this diagrammatic section through alveoli and blood capillaries shows. Hemoglobin in the red cells takes up oxygen, which passes through the alveolar wall and cell membrane. Carbon dioxide, carried in the blood as a waste product, passes the other way into the alveoli to be breathed out.

Respiratory System

The game has now switched to minimum resistance. Oxygen must be allowed to pass down a diffusion gradient, from its relatively high concentration in the air of the alveoli to its lower levels in the deoxygenated "used" blood. The smaller the barriers placed in its path, the easier is its journey. A second high-efficiency feature of the lungs is therefore the thinness of the interface for gas exchange. Each alveolus has a wall only one cell thick—and a very thin cell at that. Likewise the capillary wall is one cell thin. The distance between air and blood is about one-thousandth of a millimeter.

The blood's oxygen vehicles are red blood cells. These donut-shaped structures are tiny even for cells, but even so only just manage to squeeze single file through the pulmonary capillaries. Most of the absorbed oxygen is transported by the red cells, bound chemically to the substance hemoglobin. Red cells are packed with hemoglobin, the blood's red pigment, which acts as a "magnet" for oxygen.

As oxygen molecules latch on to hemoglobin they convert it to oxyhemoglobin, which has a much brighter, redder color. In this way the blood is transformed from its dark, red-blue deoxygenated state to the vivid scarlet of its oxygenated form. A small amount of the oxygen dissolves in the plasma (the non-cellular, liquid part of blood) and is swept away in solution.

It is said that fair exchange is no robbery, and so it is in the lungs. Carbon dioxide completes the barter as it leaves the blood, diffuses through the capillary and alveolar walls, and comes out of solution, taking its place in the air inside the alveolus.

The pulmonary capillaries reconnect to become larger and larger vessels, the pulmonary veins, which eventually return the blood to the heart. The lungs can be regarded as an intricate, intertwined network of three trees in one: airways and arteries becoming smaller, and veins getting bigger.

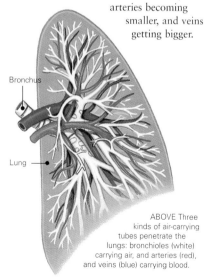

Bronchus

Lung

ABOVE Three kinds of air-carrying tubes penetrate the lungs: bronchioles (white) carrying air, and arteries (red), and veins (blue) carrying blood.

Respiratory System

THE ROLE OF SURFACTANT

Surfactant, the "natural detergent" in the alveoli, is a chemical vital to a baby's successful first breaths. A complicated mixture of protein and fat, with a substance called dipalmitoyl lecithin as its main component, it is present in the lungs through life. During expiration, the alveoli shrink and surfactant molecules are pushed closer together, so reducing the surface tension and preventing the alveoli from collapsing completely. When the alveoli swell upon inspiration, the surfactant molecules are farther apart, but the force tending to collapse the alveoli is much reduced.

BELOW The bronchi end in alveoli and each alveolus is surrounded by a network of capillary blood vessels where the arteries and veins meet. Oxygen from the incoming air passes from the alveoli into the arteries, and carbon dioxide in the venous blood passes into the alveoli.

Muscle strand

Terminal bronchiole

Alveolar capillaries

Alveolus

Recall what happens when you inflate a party balloon. At first it requires considerable effort, but once the balloon is partly expanded, it becomes easier to blow up. In the same way, the forces that tend to collapse the alveoli are greatest when the alveoli are smallest, at the end of exhalation—which is when surfactant molecules, at their closest, are most effective at lowering surface tension.

Some babies, especially those born prematurely or with inadequately developed respiratory systems, may have great trouble in breathing. The general term for such conditions is respiratory distress syndrome, or RDS. One contributing factor is lack of surfactant, which the lungs of the premature baby have not yet manufactured. Insufficient surfactant means greater surface tension in the alveoli, and the baby may not be able to overcome this. Consequently, lung function is poor and the airway linings are rapidly damaged.

DAMAGE CAUSED BY SMOKING
Tobacco smoke damages the lungs by clogging the alveoli and airways with a thick tar.... So don't smoke!

Respiratory System

HOW MUCH, AND HOW FAST?

The amount of air we can breathe, and how fast we can breathe it under various conditions of rest and activity, reveals much about the inner workings and health of the respiratory system. It therefore interests many people, from chest physicians and asthma patients to super-fit athletes looking for that extra ounce of stamina. A spirometer is an apparatus which, in its most elementary form, uses the air breathed out to displace water and make a pen trace a line across a moving chart. It measures the volume of air per unit time, which provides a measure of flow rate. More complex spirometers are linked to gas analyzers

OUR FRIENDS—THE PLANTS
As we use the oxygen in air when we breathe in and breathe out carbon dioxide, plants take in carbon dioxide and release oxygen. So encourage them to grow and talk to your green, leafy friends!

that determine the makeup of inspired and expired air.

Inspired air is usually our atmosphere, which has a relatively constant composition: nitrogen 78.08 percent, oxygen 20.95 percent, argon 0.93 percent, carbon dioxide 0.03 percent, and a miscellany of other trace gases. Expired air from quiet, restful breathing differs in two respects. The proportion of oxygen

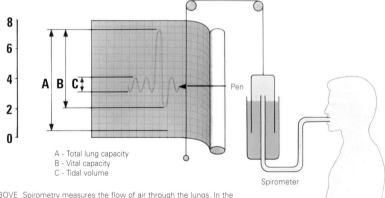

A - Total lung capacity
B - Vital capacity
C - Tidal volume

Pen

Spirometer

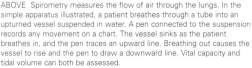

ABOVE Spirometry measures the flow of air through the lungs. In the simple apparatus illustrated, a patient breathes through a tube into an upturned vessel suspended in water. A pen connected to the suspension records any movement on a chart. The vessel sinks as the patient breathes in, and the pen traces an upward line. Breathing out causes the vessel to rise and the pen to draw a downward line. Vital capacity and tidal volume can both be assessed.

Respiratory System

has fallen to nearer 17 percent, while that of carbon dioxide has risen to around 4 percent. The respiratory system has absorbed some oxygen and "blown off" carbon dioxide. At an average of 14 breaths per minute, at a pint each, a total of 15½ pints of air are moved each minute. This enables some 3¾ pints of oxygen to enter the body and 3½ pints of carbon dioxide to be removed from it every minute.

In restful breathing, the lungs work at far below their maximum capacity. The vital statistics are as follows, for an average person. The total amount of air which the lungs could hold after the largest inhalation, that is, the total lung capacity, is some 13 pints. This volume is composed of four components.

First is the residual volume, the quantity of air that remains in the lungs even when you breathe out as hard as you can; it is air that can never be expired. It measures around 3⅓ pints. It is itself composed of two subvolumes. One is air hanging about in the contracted alveoli; the other is air remaining in the passages of the trachea, bronchi, and bronchioles. This second subvolume is the "dead space"—air that fills the airways during each respiration cycle. It is not available for the real business of gas exchange in the alveoli, hence its name.

Second is the expiratory reserve volume. This is "extra" air you can manage to force out of your lungs after already having exhaled a normal breath. Its volume averages 2⅓ pints.

Third is the tidal volume—the air shifted in and out during normal, restful breathing. As mentioned previously, it is around 1 pint, but increases rapidly with exercise.

Fourth is the inspiratory reserve volume. You inhale for an average breath, but then go on inhaling as far as you can. The amount of air drawn in, over and above the tidal volume, makes up this fourth component. It averages 6½ pints. Add the four together: the answer, and total lung capacity, is 13 pints.

There is not a lot that can be done about the lungs' residual volume. But expiratory reserve plus tidal plus inspiratory reserve volumes—together constituting the lungs' vital capacity—often provides a useful pointer to general health. And in some respiratory illnesses, termed restrictive lung diseases, the vital capacity is reduced.

Another gadget, much simpler and cheaper than the spirometer, is a peak flow meter. It gauges the fastest (peak) rate at which air can be blown out of the mouth, and hence up from the lungs. The meter is a useful ready reckoner in respiratory illnesses, for instance, in assessing the severity of an asthma attack and the efficacy of treatment.

BRAIN POWER
You don't have to tell your lungs to keep breathing—your brain does it automatically for you.

Respiratory System

LUNG DISORDERS

Most lung disorders can be classified according to the pattern of functional abnormality: the effect on the lungs, rather than the cause. In obstructive lung conditions such as asthma and emphysema, the major problem is that the air flow through the conducting airways is obstructed or limited in some way. Sufferers undergoing pulmonary function tests inhale or exhale air at a rate much lower than a comparable healthy person. The obstructed airflow may result in audible wheezing, as in asthma.

The airway obstruction may be a partial plugging from build-up of mucus in the air passages, as in patients with chronic bronchitis. It may be disruption and expansion of the fibrous supporting tissue around the airway, which compresses the bore of the passage, as in emphysema. It may be a thickening of the airway wall, as in some types of "dust disease" (pneumoconiosis, caused by long-term inhalation of damaging dust particles). Or it may be contraction of the smooth muscle bands encircling the bronchial walls; asthma affects the bronchioles in this way.

In restrictive lung disorders, the problem is a lowered total lung capacity. In lung function tests, when patients with restrictive diseases are asked to breathe in and fill their lungs completely, they are unable to take in an adequate volume of air compared to healthy people of the same weight and height. There is no obstruction to airflow, unless obstructive disease is also present (as happens in emphysema).

Restriction of lung capacity can be caused by any of three factors: disease in the lung itself, problems with the pleurae

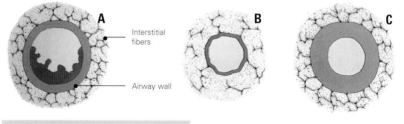

A — Interstitial fibers
A — Airway wall

PASSIVE SMOKING
Secondhand smoking from somebody else's cigarette can affect you and pollute the atmosphere, and is damaging to your lungs as well as the smoker.

Airways become blocked when obstructive lung disease strikes. This may happen in one of three ways. Excessive mucus may block the airway (A); loss of traction in the surrounding interstitial fibers may cause narrowing (B); or the airway wall may thicken and restrict airflow (C). Whatever the cause, the result is a reduced flow of air and impaired efficiency of respiration.

Respiratory System

covering the lungs, and malfunction of the chest wall or muscles of respiration.

In the lung itself, diseases such as sarcoidosis and asbestosis affect the interstitial tissue—that is, the "ground matter" of elastic fibers and supporting connective tissue, around and between the airways and alveoli. The lungs lose their elasticity, or compliance, and become stiff and difficult to inflate to the required degree. Paradoxically, the airways may not be involved in the disease process itself but they may become dilated, or wider than usual. These dilated small airways, surrounded by scarred, thickened connective tissue, have been likened to a honeycomb in appearance, hence the term "honeycomb lung" to describe severe interstitial restrictive lung disease.

Pleural problems can also lead to restriction of lung volume. The most striking example is pneumothorax, when air from outside enters the space between the two layers of pleural membrane around each lung. The air destroys the "seal" by which the breathing muscles and rib cage stretch the lungs upon inhalation. The lung, no longer under tension, collapses. Spontaneous pneumothorax is usually caused by the rupture of a small air space in the top of the lung. It is five times more common in men than women, and particularly affects young, tall thin men, and more frequently the right rather than the left lung. Provided the leak is short-lived and soon sealed, the air in the pleural space is gradually absorbed over ensuing weeks

and the lung expands back to normal.

If the chest wall suffers trauma and is pierced, perhaps by a knife or a bullet, or as the result of an automobile accident, the situation may be more serious. The lung collapses as air floods the pleural space. With each breath, air is sucked in and out through the wall rather than up and down the trachea; the lungs stay shrunken. Since each lung has a separate pleura, damage to one does not necessarily affect the other. But air entering through such a "sucking chest wound" may build up in the chest, compressing and threatening the other lung and the heart. As the victim struggles harder for breath, the condition worsens. Emergency treatment is vital, firstly to prevent inflow and accumulation of air, and secondly to remove the indrawn air from the pleural space and allow the lung to reinflate.

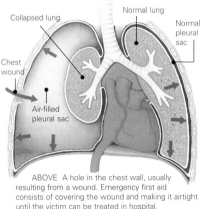

Collapsed lung

Normal lung

Normal pleural sac

Chest wound

Air-filled pleural sac

ABOVE A hole in the chest wall, usually resulting from a wound. Emergency first aid consists of covering the wound and making it airtight until the victim can be treated in hospital.

NERVOUS SYSTEM

Silently and ceaselessly, at the height of human activity and during the depths of sleep, our nerves relentlessly transmit impulses to and fro from the brain. Every movement we make, from an unconscious twitch of an eyelid to the delicate actions of driving a motor car, and every sensation we feel—even our dreams —depend on the workings of the nervous system.

This complex and sophisticated network extends throughout the body. It is constantly gathering information of all kinds and transmitting commands to stimulate muscles and other organs into activity.

The main parts of this vital system are the brain, the site of sense and sensation; the spinal cord, the main trunk route of the communications network; and the nerves themselves, a branching network whose roots lead into and out of the spinal cord. It is so intricate that the exact workings of each of these elements is still not fully known. Nevertheless it is possible to obtain a broad picture of how the complete system controls the body. Higher functions—those involving memory, comparison, and decision making— take place in the brain, while the rest of the

nervous system carries sensory information to the brain and commands from it to control the body's movements and reactions.

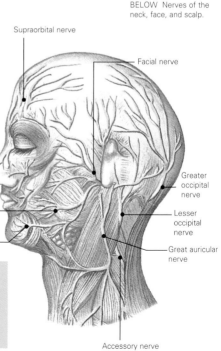

BELOW Nerves of the neck, face, and scalp.

Supraorbital nerve

Facial nerve

Greater occipital nerve

Lesser occipital nerve

Buccal nerve

Great auricular nerve

Mental nerve

Accessory nerve

HUMANS
Human beings have the most highly evolved nervous system of any living creature.

Nervous System

SYSTEMS WITHIN SYSTEMS

Unlike the body's blood vessels or lymph canals, the nerves do not form a single system, rather there are several interrelated systems. Some of these are physically separate; others differ only in function. Together the brain and spinal cord make up the central nervous system. The rest of the network is the peripheral nervous system; physically part of it—but with its own specific functions—is the autonomic nervous system. This is responsible for those body functions not under conscious control—
the unfailing beating of the heart and the automatic operation of the kidneys and the digestive system. The harmonious, smooth operation of these vital body functions is achieved by further dividing the autonomic nervous system into parasympathetic and sympathetic systems. These have broadly opposing actions and so provide checks on each other, working together to achieve a balance.

Like all other tissues in the body, from bone to blood, nerves are made up of cells. But nerve cells, termed neurons, are remarkable in many respects and include some of the largest— or at least longest—cells in the body, some being more than a foot long. The nervous system operates by means of minute electrical signals, or impulses, traveling along the length of the nerve cells. Acting at high speed it processes information from the sensory nerves, and initiates any required actions in a fraction of a second. Nerve impulses travel at up to 250 miles per hour.

THE LONGEST NERVE
One of the vital nerves that arise directly from the brain, the vagus,p is among the longest nerves in the body. It carries messages from the brain to stimulate the secretion of digestive enzymes and activate the smooth muscles that propel partly digested food along the gut.

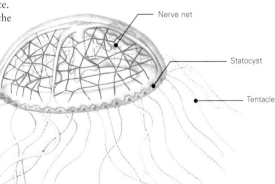

Nerve net

Statocyst

Tentacle

RIGHT Jellyfish have a diffuse network of nerve cells in contact with each other.

Nervous System

A BRANCHING NETWORK

An anatomy teacher describing the nervous system has a similar problem to a wireman trying to explain how a telephone exchange works—it is extremely complex and has no real beginning or end. Within the body the nerves branch out like telephone lines from an exchange. They run to every extremity of the body, from the soles of the feet to the top of the scalp, and from the surface of the skin to the central organs of heart, liver, and lungs. These nerves are actually single cells which have the highly specialized function of carrying information from one part of the body to another. Most are bunched together like the strands of a rope and form a cord. Individual nerves may be as small as one six-thousandth of an inch thick. They look creamy white, though this color fades as the nerve bundles split up.

Nerve cells have the same basic structure as all the other body cells, with a surrounding membrane containing the nucleus and cytoplasm. But they have a very special shape. Consider a typical motor nerve carrying instructions from the brain to a muscle. At one end there is a tuft of short rootlike projections or dendrites. At the other is a long thin projection, the axon, which may subdivide up to 150 times and be attached to 150 separate muscle fibers.

Although nerve cells may be thinner than a hair from your head, they may also be surprisingly long. For example, in

CENTRAL SYSTEM
The brain and spinal cord are known as the central nervous system. The smaller nerves that come out of the brain and spinal cord are called the peripheral nervous system.

BELOW This sponge, for example, has no nerves at all, and relies for information on "messenger" substances released by ameboid cells in its walls.

Ameboid cells

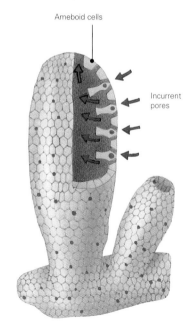

Incurrent pores

Nervous System

a typical human the nerve running from the base of the spine to the tip of a toe is around three feet long. However, many other axons are only a fraction of an inch long. Most nerves act as links in a chain of nerve cells rather than connecting directly to a muscle. In such a chain each axon is in near contact with the dendrites of the next cell, but there is a tiny gap, known as a synapse, between them. Nerve impulses must somehow jump this gap and they do this with the help of chemical messengers, which are known as neurotransmitters.

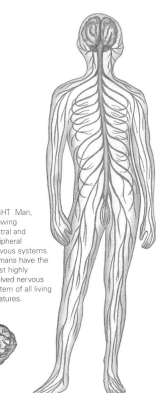

RIGHT Man, showing central and peripheral nervous systems. Humans have the most highly evolved nervous system of all living creatures.

BELOW Earthworms have a similar arrangement to flatworms, consisting of lateral nerves branching off in each segment of the body.

Paired nerves

Ventral nerve cord

Cerebal ganglion

EVEN THE BONES
Nerves reach every part of the body, even the bones.

Nervous System

A CHAIN REACTION

When an electrical nerve signal reaches the end of an axon, it triggers the production of a neurotransmitter chemical from tiny secretary cups, or vesicles. The neurotransmitter then quickly diffuses across the synapse and excites the dendrites of the next cell to induce a new electrical signal in their cell, and so on along the chain. This knock-on effect happens in a fraction of the time it takes to explain it. At each synapse there is some resistance to the transmission, so the nervous message gradually becomes weaker as it moves along the nerve chain.

Although the synaptic gaps appear to hold up the smooth flow of nerve signals, they have the vital function of allowing more than one nerve cell to influence the next. And cells can be subjected to many different influences, some of which may stimulate the cell while others "switch off" or inhibit it. The cell's response then depends on the balance between the various stimuli, so the multiple path action allows the system greater precision and control.

Nerve signals are one-way traffic: impulses can flow only from the axon of one cell to the dendrite of another, and reverse flows are impossible. So when you start walking, for example, the control impulses travel from your brain, down the spinal cord, and along nerves to the leg muscles. These "command" impulses are transmitted along motor neurons, which are also known as efferent fibers because they carry messages away from the brain. But if you stub your toe, the sensations from your foot are carried back to the central nervous system by another set of nerve cells. These sensory neurons start from the sense organs and are described as afferent because their information travels toward the spine and brain. Both types of nerve may run in the same nerve bundle and they look much alike.

BELOW In insects the cerebral ganglion forms a primitive brain that controls various activities.

Cerebral ganglia

Ventral nerve cord

Thoracic ganglia

A NEURON
A neuron is a nerve cell. The brain is made up of approximately 100 billion of these nerve cells.

Nervous System

GROWING COMPLEXITY

Because the human nervous system is so complex, it is difficult to analyze and understand. In contrast, many animals have much simpler nervous systems, and studying them can help give useful insights into how the human system works.

Very simple creatures such as amebas and sponges manage without any nervous system at all. The most primitive animals with nerves are sea anemones, jellyfish, and corals. Their body cells are arranged in outer and inner layers, with a "nerve set" at the base of the outer layer. The net consists of interconnecting nerve cells, with two or three axons from each cell linking to the axons of other cells. As with vertebrate animals the axons do not fuse with each other—a tiny synapse is left that the nerve impulse has to jump across.

But with these simple animals all the axons are similar and there are no branching dendrites. Impulses can travel in either direction along the nerve cell, so any stimulation radiates across the entire body through the network. There are faint traces of nerve specialization in some species of sea anemone; they have one or two clearly marked conduction paths that let the impulses travel more quickly.

In a sea anemone the nerve cells are concentrated around the mouth region, but there is no brain. All higher animals have a distinct head region and most of

them can move around. Movement makes it easier for an animal to exploit its environment but raises a different set of problems. Having a head—which leads the way during movement—allows the development of specialized sense organs to gather the information that helps solve these problems.

CONTROL CENTER
Being the control center for your whole body, the nervous system is made up of your brain, spinal cord, and an enormous network of nerves that run throughout your entire body.

BELOW Nerves of the hand.

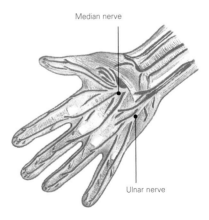

Median nerve

Ulnar nerve

Nervous System

Flatworms show early stages in the development of such a system with rudimentary eyes near one end of their bodies. Although the eyes do little more than respond to light (which the flatworms shun), there is a definite concentration of nerve cells around them. These cells have a simple brainlike function that is recognized by calling them the cerebral ganglia. Two thick bands of nerve cells run back from the cerebral ganglia to form distinct nerve cords. This is an arrangement that is used by almost all of the more advanced animal groups.

TELEPHONE WIRES
The nerves send messages to and from the brain in just the same way telephone wires work.

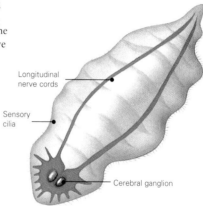

Longitudinal nerve cords

Sensory cilia

Cerebral ganglion

ABOVE Flatworm, showing its nervous system. They have a centralized system with cerebral ganglia and a pair of nerve cords.

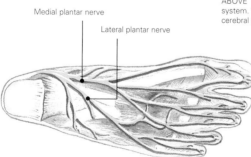

Medial plantar nerve

Lateral plantar nerve

LEFT Nerves of the foot.

Nervous System

THE BONUS OF THE BRAIN

Further evolution resulted in "modular" animals made up of a number of repeating sections or segments, like beads on a string. Earthworms and leeches show this clearly—the segments can be seen on the outside of the body. Each segment contains a basic set of vital organs, and each contains a nerve cell ganglion and

paired lateral nerves, with the ganglia of adjacent segments connected to each other. In addition there are three giant fibers that run the whole length of the worm. In the event of a violent stimulus these fibers—which bypass the ganglia— give rapid transmission of nerve impulses, and enable a rapid reaction.

TO AND FROM NERVES
Sensory nerves send messages to the brain, motor nerves carry messages back from the brain to all the muscles and glands in your body.

BELOW A single nerve fiber consists of a chain of neurons, or nerve cells. Each has a cell body containing a nucleus and bearing rootlike dendrites. An axon extends from a "hillock" on the cell body, and is wrapped around with Schwann cells separated by gaps at the nodes of Ranvier. Terminal buttons at the end of the axon link it to other nerve cells.

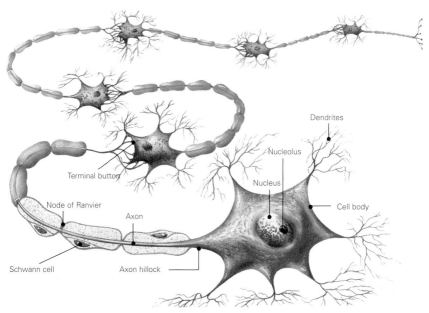

Dendrites

Nucleolus

Nucleus

Cell body

Terminal button

Node of Ranvier

Axon

Schwann cell

Axon hillock

Nervous System

In the head of the earthworm there is a fairly well-formed cerebral ganglion, and together with another ganglion under the gullet this forms a workable brain. But the brain is not in total command of the body; its main function is to integrate the activities of the separate ganglia and of the body as a whole. If an earthworm is cut into two pieces the rear piece—no longer connected to the brain—remains as active as the front portion.

ELECTRIC PULSES
When a neuron is stimulated, whether it be by touch, sound, or some other message, it begins to actually generate a tiny electrical pulse.

Nerve cell body

BELOW Schwann cells form the insulating sheath of myelin that encircles a peripheral nerve axon. The diagram shows how the sheath gradually thickens on a developing nerve by coiling around it. It forms a bulge to house the large nucleus of the Schwann cell.

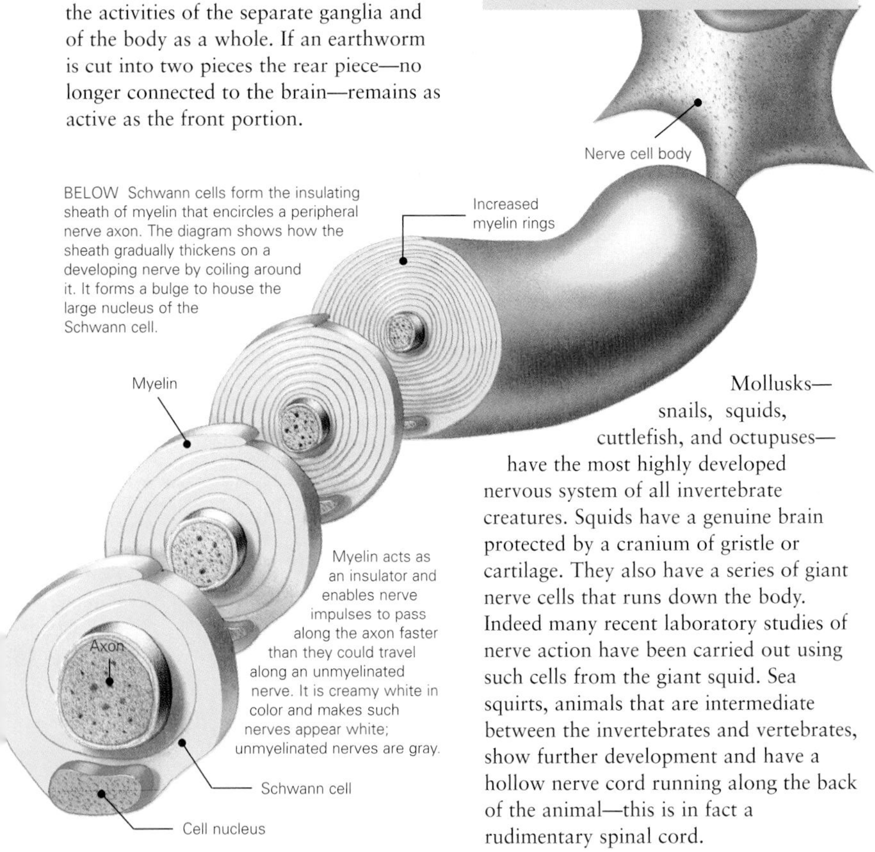

Increased myelin rings

Myelin

Myelin acts as an insulator and enables nerve impulses to pass along the axon faster than they could travel along an unmyelinated nerve. It is creamy white in color and makes such nerves appear white; unmyelinated nerves are gray.

Axon

Schwann cell

Cell nucleus

Mollusks—snails, squids, cuttlefish, and octopuses—have the most highly developed nervous system of all invertebrate creatures. Squids have a genuine brain protected by a cranium of gristle or cartilage. They also have a series of giant nerve cells that runs down the body. Indeed many recent laboratory studies of nerve action have been carried out using such cells from the giant squid. Sea squirts, animals that are intermediate between the invertebrates and vertebrates, show further development and have a hollow nerve cord running along the back of the animal—this is in fact a rudimentary spinal cord.

Nervous System

THE INTRICATE NETWORK

The intricate network of the nervous system extends from the brain and spinal cord to reach every part of the body. The nerves of the peripheral system are involved in two-way traffic. Stimulation of sensory organs and receptors—which provide us with our senses of sight, hearing, taste, smell, and touch—triggers them to send messages along nerves to the brain, where they are interpreted. The brain sends back instructions to the organs and tissues along the cranial nerves or via the spinal cord, enabling us to respond almost simultaneously.

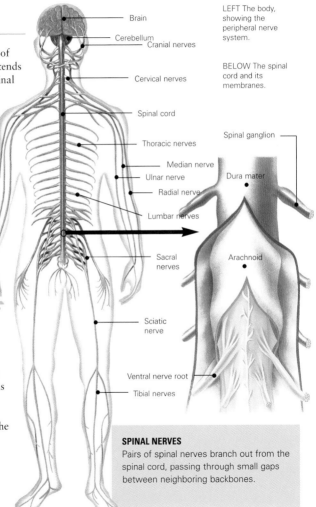

Brain
Cerebellum
Cranial nerves
Cervical nerves
Spinal cord
Thoracic nerves
Median nerve
Ulnar nerve
Radial nerve
Lumbar nerves
Sacral nerves
Sciatic nerve
Ventral nerve root
Tibial nerves

Spinal ganglion
Dura mater
Arachnoid

LEFT The body, showing the peripheral nerve system.

BELOW The spinal cord and its membranes.

SPINAL NERVES
Pairs of spinal nerves branch out from the spinal cord, passing through small gaps between neighboring backbones.

Nervous System

THE EXTENSIVE NERVE NETWORK

The lungs, liver, digestive organs, and most of the other vital systems are concentrated in a particular part of the body. But the peripheral nervous system is widespread and diffuse. Nerves travel out from the spinal cord, branching repeatedly to form a network that extends throughout the body. Ultrasensitive regions such as the fingertips and lips have a high concentration of nerve endings, whereas less sensitive regions such as the back have fewer nerves. Depending on their location, the sensory nerves can end in receptors that respond to touch, temperature, or certain chemicals. Some particularly sensitive regions, such as the tongue, have large numbers of all three types of receptors.

Most of the nerves of the peripheral system branch out from the spinal cord like branches from the trunk of a tree. Collections of nerve cells (the ganglia) just outside the spinal cord act as sorting centers and help channel the information flow and any responses. Spinal nerves control all bodily movements, and this is why serious back injury that damages the spinal cord can result in paralysis. A few nerves, the cranial nerves, run directly from the brain to the face, teeth, and mouth. Below the neck the only major cranial nerve is the vagus—which belongs to the parasympathetic part of the automatic nervous system—extending all the way to the intestines.

Most nerve cells have a single axon and a large number of dendrites. These are around one-twentieth of an inch long and branch repeatedly to give the neuron a structure resembling a bush. The tips of the dendrites terminate in thousands of smaller projections known as dendrite spines. Impulses arrive from other nerve cells through the spines, so that each neuron can receive messages from a number of others. All the messages are integrated in the cell body; if they result in the creation of an additional impulse, this is transmitted along the axon.

An axon generally emerges from the cell body at a raised region called the axon hillock. In humans the axon varies from about one ten-thousandth to one-thousandth of an inch in diameter, and may be as short as one-hundredth of an inch or as long as three feet.

Usually the axon does not branch along its length, but where it does the collaterals (branches) normally run off at right angles. Branching does occur at the ends, but not to the same extent as with dendrites, and each branch normally ends in a knoblike swelling—the terminal bouton or button. The swellings make contact with the dendrites, cell body, or axon of other cells, or alternatively they end at muscle fibers.

SPEED OF TRAVEL
Information travels between 0.5 meters/sec and 120 meters/sec through the nervous system.

Nervous System

SUPPORTING GLUE

Individual neurons are supported and held together by a mass of neuroglia (literally "nerve glue"), which is made up of large numbers of separate cells. The glial cells also form a barrier between the bloodstream and the nervous tissue, only letting through essential substances such as oxygen, water, and some sugars. There is

SYNAPSES
Within the human brain, it has been estimated, that there are:
1,000,000,000,000,000 synapses, which is equal to about half a billion synapses per cubic millimeter.

only one sort of glial cell—a Schwann cell—in the peripheral nervous system.

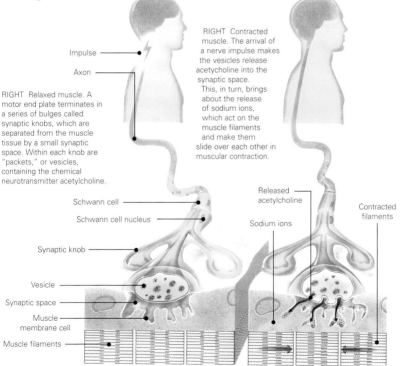

Impulse

Axon

RIGHT Relaxed muscle. A motor end plate terminates in a series of bulges called synaptic knobs, which are separated from the muscle tissue by a small synaptic space. Within each knob are "packets," or vesicles, containing the chemical neurotransmitter acetylcholine.

RIGHT Contracted muscle. The arrival of a nerve impulse makes the vesicles release acetycholine into the synaptic space. This, in turn, brings about the release of sodium ions, which act on the muscle filaments and make them slide over each other in muscular contraction.

Schwann cell

Schwann cell nucleus

Synaptic knob

Vesicle

Synaptic space

Muscle membrane cell

Muscle filaments

Released acetylcholine

Sodium ions

Contracted filaments

Nervous System

Passing on the Message

Information carried along one nerve is transferred to the next neuron, or to a muscle, through the synapses. The nerve impulse jumps the gap with the help of a "ferryboat" in the form of a chemical messenger substance. A typical synapse is made up of several discrete parts: it has a "before" presynaptic region, such as the terminal bouton or an axon of one neuron; the synaptic cleft or gap itself; and an "after" or postsynaptic region such as the dendritic spines of another neuron. Motor nerves end on muscle fibers in elongated structures called motor end plates. These plates are longer than

> **THE SPINAL CORD**
> On average the spinal cord in men is about 17.7 in. (45 cm) long and in women it is around 16.9 in. (43 cm) long.

the presynaptic boutons, but the general structure of the synapse is much the same.

The presynaptic region may be swollen to form a presynaptic knob containing many "packages" of one or more different neurotransmitter chemicals. The synaptic cleft itself is very small and is filled with a sticky material which glues the presynaptic and postsynaptic regions together.

The eyes send sensory messages to the brain about the position of the glass.

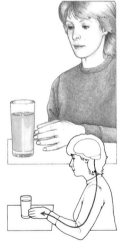

The brain sends motor impulses to the muscles of the arm to move it toward the glass.

Touch receptors in the hand detect the glass and inform the brain about contact.

Nervous System

When the impulse in an axon reaches the presynaptic terminal, it triggers the release of the neurotransmitter acetylcholine. This chemical diffuses across the synaptic gap and joins with receptor molecules on the postsynaptic membrane. There it sets off an electrochemical reaction (depolarization) which excites the next nerve cell or the muscle. The process is not quite instantaneous because it takes time for the neurotransmitter to diffuse across the synapse. There is a delay of around half a thousandth of a second between the impulse reaching the presynaptic terminal of one axon and the corresponding action potential starting in the next nerve cell or muscle.

THE LENGTH OF A NEURON
Some neurons are less than a millimeter long. Others can be as long as a meter or more, for example the axon of a motor neuron in the spinal cord that innervates a muscle in the foot measures approximately 3 ft. (1 m).

After release, the acetylcholine remains active for only one or two thousandths of a second before it is broken down by enzyme action, so ensuring that its effect is short lived. But the inactivated chemical is not wasted; it is taken up by the presynaptic membrane, where it is converted back into reusable acetylcholine in the presynaptic knobs.

Motor impulses from the brain make muscles grip the glass and raise the arm.

Sensory impulses from the eyes and arm inform the brain about the arm's position.

The lips send sensory impulses to inform the brain that the glass has reached the mouth.

Nervous System

SIGNALS THAT WORK TOGETHER

When they trigger the next nerve cell, which they do by depolarizing the postsynaptic membrane, neurotransmitters have a positive effect. But synapses can also have a negative, inhibiting effect when the arrival of neurotransmitters causes an increase in the synaptic membrane polarization. This makes subsequent depolarization more difficult, and it may slow down the impulse rate if the nerve cell is already active. As a result, a synapse may be either excitory or inhibitory, depending on where it is and on the neurotransmitter substance. For instance, acetylcholine is excitory at nerve-muscle junctions but inhibitory in some parts of the autonomic nervous system.

A single nerve cell, such as a motor cell in the ventral horn of the spinal cord, may have many hundreds of thousands of synaptic contacts—all of which may have excitory or inhibitory effects. Even when the arrival of a single impulse at an excitory synapse is not powerful enough to depolarize the cell enough to generate an impulse a following impulse may have an add-on effect and do so; this cumulative effect is called temporal summation.

HOW HEAVY IS THE BRAIN?
In humans, a newborn baby's brain will weigh about 12–14 oz. (350–400 g), and a fully grown adult's brain weighs between 2.9–3 lbs. (1300–1400 g).

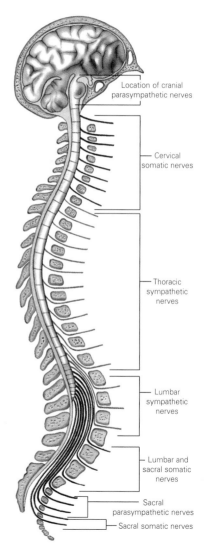

Location of cranial parasympathetic nerves

Cervical somatic nerves

Thoracic sympathetic nerves

Lumbar sympathetic nerves

Lumbar and sacral somatic nerves

Sacral parasympathetic nerves

Sacral somatic nerves

Nervous System

The phenomenon known as spatial summation is similar, but by contrast to temporal summation, it occurs when two synapses that are separated in space (on the same nerve cell) are excited at the same time.

Impulses at synapses with opposite effects can, however, cancel out each other. Or a strong impulse may override a weaker one of an opposite type. The overall response of a nerve cell to the many signals it receives therefore depends on the integrated effect of all its active synapses. This means that the response of a single nerve cell may depend on the input from hundreds of thousands of other neurons. The end result of all this activity may be, for example, a tiny, precisely controlled body movement.

LEFT and RIGHT The autonomic nervous system consists of twin parts, known as the sympathetic system (red) and the parasympathetic system (blue). All the sympathetic nerves arise from the spinal cord, from 15 segments that make up the thoracic and upper lumbar regions. The parasympathetic nerves include four cranial nerves, directly from the brain, and eight spinal segments from the lower lumbar and sacral regions. The other eight cranial nerves are shown in yellow on the diagram, and the somatic nerves, from the cervical segments of the spine, are shown in black.

LEFT Underside of brain showing the autonomic nervous system.

NEURONS DIFFER FROM OTHER CELLS
Neurons differ from other cells in the body as they communicate with each other through an electrochemical process.

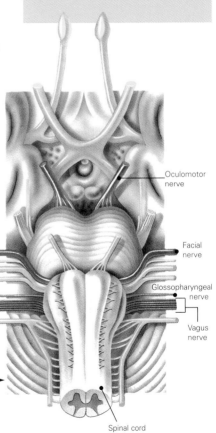

Oculomotor nerve

Facial nerve

Glossopharyngeal nerve

Vagus nerve

Spinal cord

Nervous System

RAPID RESPONSE SYSTEM

The brain does not control all of the actions of the peripheral nervous system. When a rapid reaction is necessary, say in an emergency situation, the spinal cord can trigger the appropriate response. A typical example is the way you jerk your hand back after touching a hot surface. Sensory messages from the fingers travel up the afferent fibers to let the brain know that something is causing pain. The brain responds at once by sending messages to move the fingers. But at the same time the messages act directly through a reflex arc in the spinal cord to instantly trigger motor impulses that jerk the hand away. This reflex action moves the hand a few thousandths of a second before brain withdrawal impulses arrive, and so helps minimize any damage.

A physician can make a simple check on how some of the spinal reflexes are working by using the simple stretch reflex. This safeguards muscles from rupture by making them suddenly contract when they are stretched. Impulses from the muscle travel up the sensory fibers to the spinal cord and pass across it to produce a motor impulse that

contracts the muscle. In a typical test the physician taps the tendon below the knee to stretch the muscle, and the leg gives a sharp jerk as the muscle contracts.

The stretch reflex is the simplest of the spinal reflexes because the impulse passes

LEFT and OPPOSITE
A reflex is a defensive mechanism. It results in an almost instantaneous muscular reaction to an emergency stimulus, and does not directly involve the brain. In the example illustrated, the immediate response is a pain signal (green) sent by sensory nerves toward the brain. But it is intercepted at the base of the spine, which returns an instant command (red) along the motor neurons to the muscles of the leg, which contract to pull the foot away from danger.

TWO FACTS ABOUT NEURONS
1. Neurons are surrounded by a membrane.
2. Neurons have a nucleus that contains genes.

Nervous System

directly from the sensory to the motor nerve. All the other spinal reflexes use additional nerve cells—known as interneurons—to transfer the signal. And in some reflexes the interneurons can also come into play to balance other kinds of muscle contractions.

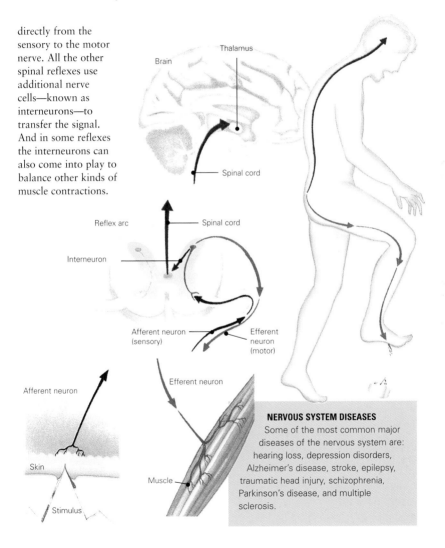

Brain

Thalamus

Spinal cord

Reflex arc

Spinal cord

Interneuron

Afferent neuron (sensory)

Efferent neuron (motor)

Efferent neuron

Afferent neuron

Skin

Stimulus

Muscle

NERVOUS SYSTEM DISEASES

Some of the most common major diseases of the nervous system are: hearing loss, depression disorders, Alzheimer's disease, stroke, epilepsy, traumatic head injury, schizophrenia, Parkinson's disease, and multiple sclerosis.

Nervous System

LEVELS OF DISCRIMINATION

Nerve sensations register as pain only when the stimulation rises above a specific level. This "threshold" effect is due to the action of an ingenious gate mechanism, which modulates the signals traveling up the spinal cord. Pain information is transmitted by spinal cord transmission (or T) cells, activated by large or small nerve fibers running from the source of the stimulation. Within the substantial gelatinosa (SG) of the dorsal spinal cord there are other cells that act to inhibit the T cell transmission. These cells are activated by large-fiber input to the spinal cord, but they are also inhibited by the C fibers—the small pain fibers.

When no pain is present, the fibers concerned with touch and position sensations are activated together with the SG cell. This inhibits the T cell, closes the gate, and there is no pain perception.

As soon as a painful stimulus occurs—say you step on a sharp rock—the activity of the small C fibers increases to a level where it overcomes the inhibition of the large-fiber activity of the SG cell. Once this happens the gate is open and a pain stimulus travels up the spinal cord to tell your brain that your foot is injured. The immediate reaction is to rub the painful area. The rubbing action increases the level of touch sensation, which augments large-fiber activity and so, in turn, activates the SG cells to close the gate. So rubbing a bruise does not make it better, but it does make it hurt less.

A GANGLION

In the case of the autonomic nervous system, a ganglion (plural, ganglia) refers to a group of nerve cell bodies. Sympathetic and parasympathetic ganglia are located along the spinal cord and in the body cavities.

LEFT Pain threshold map viewed from front of body.

Nervous System

RIGHT Viewed from the side, the skin's sensitivity, and hence the threshold to pain, varies over the surface of the body. It is lowest on the feet and highest on the head, but varies according to the age and vitality of the individual. Pain threshold can also depend on circumstances; a person used to hard manual labor, or a soldier in battle, has a much higher threshold than somebody who sits behind a desk all day.

CONQUEST OF PAIN

The sensation of pain is a natural warning that tells you something is wrong with the body: perhaps your hand is too close to a fire or the food you have just put into your mouth is bad. But once the warning has been heeded and action taken—you move your arm away from danger or spit out the offending mouthful— the persistence of pain is an ongoing problem, particularly if, like the sufferer from a headache or rheumatism, there is little or nothing you can do about it. The conquest of pain therefore ranks among the greatest achievements of modern medicine.

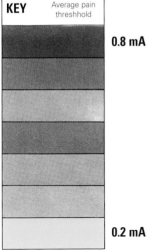

KEY	Average pain threshhold
	0.8 mA
	0.2 mA

STRONGER THAN MEN
In general, women can withstand predictable physical pain better than men.

DIGESTIVE SYSTEM

Most people enjoy eating, and food is a popular topic of conversation. It is also a popular topic with writers and philosophers. François Rabelais, sixteenth-century French cleric and renowned scholar, in his epic and satirical Gargantua et Pantagruel, penned in anticipation: "I drink for the thirst to come . . . Appetite comes with eating . . ."

At the other end of the spectrum, it is difficult to argue with the deadpan realism of Ludwig Feuerbach, nineteenth-century German philosophical materialist: "A man is what he eats."

On the outside, we call it food. It may be a crust of bread or a lavish 15-course banquet, but the digestion shows no favors. A healthy human eats about half a ton of it each year. If this food is to be of value to the body, rather than merely pleasing to the taste buds, then it must be digested. The system that performs this complex cascade of chemistry is in essence a tube, more than 25 ft. (7.6 m) long.

The digestive tube is twisted, swollen, and shaped along its length into several distinct regions: mouth, pharynx (throat), esophagus (gullet), stomach, ileum, colon (small and large intestines), rectum, and anus. Food passing along the system is mixed with a staggering laboratory of chemicals, which break it down by means of a multitudinous series of interlocking reactions into small molecular units. These can be absorbed into the body, chiefly into the blood and lymph systems.

On the inside, the nutritive molecules succumb to a variety of possible fates, depending on the body's long-term needs and the necessities of the moment. Some molecules act as energy sources that power the biochemical gyrations of every cell; others become building-blocks, for construction of tissue during growth, or in reconstruction of repaired or worn-out tissues. In times of plenty, certain molecules may simply be set aside until times of shortage—one of the body's many forms of life insurance.

Not all we consume is digestible. Wastes remain, and the system must package and dispose of them in an efficient way. It may be the less glamorous end of the tale, but no less important for the system's functioning—and so just as vital for good health.

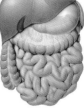

THE DIGESTIVE SYSTEM
The digestive system is composed of the digestive or alimentary tube and accessory digestive organs.

Digestive System

ESSENTIAL FUELS

A striking feature of human digestion is the wide range of foodstuffs on which it can survive and prosper. In zoological terms, humans are perhaps best described as omnivores with a definite leaning toward herbivory. Our adaptable guts, as well as our inventive brains, have allowed this species of large primate to spread across the globe and utilize for food whichever edible plants and animals were present in the region. Today, a significant proportion of the world's five billion human digestive systems exist on a relatively, restricted, localized diet—most people eat what they grow. In the more affluent countries, people are able to sample foodstuffs imported from various places around the world.

The three principal constituents of food are proteins, carbohydrates, and fats. The first group of substances is used essentially for construction and repair, while the second fuels the body's processes and functions; the third does both, building and fueling. In addition the body must have vitamins, which are essential for normal growth and development, and minerals, which are substances required in tiny amounts to assist in many body processes such as nerve and muscle function.

Paradoxically, the diet as a whole should also contain material that is not actually digestible, such as the cellulose in plants. It may be called "fiber," "roughage," "bulk," and various other names. The usefulness of fiber in the diet is slowly becoming understood, in that it seems to contribute physically and chemically to a healthily efficient digestive system. Its absence is currently linked to a variety of intestinal ailments, from constipation to colonic cancer.

The final and central dietary requirement forms the medium in which all else happens. It contains no energy, no building blocks, no vitamins, no fiber. It is plain water.

Unsaturated fat

Saturated fat

LEFT One of the three major categories of macronutrients in food is fats, which include both saturated and unsaturated fat.

IN THE MOUTH
In the mouth, foodstuffs are broken down mechanically by chewing and saliva is added as a lubricant.

Digestive System

BUILDING WITH PROTEINS

About one-sixth of your body weight consists of proteins, which make up the main part of tissues such as muscles, skin, hair, and nails. All proteins contain the element nitrogen, as well as the carbon, hydrogen, and oxygen found in many other organic molecules. They are big molecules: one protein molecule may contain thousands of atoms, and the larger ones can be seen individually using today's advanced electron microscopes. There are two main types of protein, structural and functional.

Structural proteins are the body's building-blocks, scaffolding, girders, cladding plates, and tiles. Functional proteins are the enzymes, which are the chemical handlers and regulators—the construction workers themselves. All the thousands of processes upon which human life depends are controlled by enzymes, and all enzymes are functional proteins. (Enzymes are especially prevalent in the chemistry of digestion, where they act as organic catalysts to break up large food molecules into ones small enough to be absorbed.)

Each protein eaten in food is made of a chain of subunits, called amino acids, strung like beads on a necklace. A small protein may have one or two dozen amino acids; a large one has many hundreds. In the entire animal kingdom there are only about 20 different amino acids. Yet these can be linked and repeated in endless combinations, in the way that the 26 letters of the English alphabet may be shuffled to make the tens of thousands of words in common usage. Each protein, like each word, has individual "meaning" in its structure.

Glucose

ABOVE
The simplest carbohydrate, and the subunit into which more complicated ones are broken down during digestion, is the blood sugar glucose. The same three elements can combine in other ways to form saturated and unsaturated fats (see previous page).

Hydrogen   Amino acid side chain   Nitrogen   Carbon   Oxygen

THE BASIC DIGESTIVE SYSTEM
The digestive system is basically a series of hollow organs joined in a long, twisting tube from the mouth to the anus.

Digestive System

FUEL FOR THE FURNACE

Sugar, and especially glucose sugar, is the body's preferred energy source. Chemically, carbohydrates are combinations of many individual sugar molecules, the smallest and simplest of which is glucose. The name "carbohydrate" derives from its constituents: carbon, hydrogen, and oxygen.

Enormous chains of sugars congregate to make up those very large carbohydrates, the plant starches, which are eaten in the form of rice, potato, grains, and many other plant parts. During digestion, these huge chains are snipped up into smaller ones that can be absorbed and either burned, to fuel cellular metabolism, or reassembled and stored in the energy closets in the form of glycogen ("animal starch").

WHY IS DIGESTION IMPORTANT?
Our food and drink must be changed into smaller molecules of nutrients before they can be absorbed into the blood and carried to cells throughout the body.

RIGHT Far more complicated in molecular structure, proteins additionally contain nitrogen, either as linking atoms or in amino acid side chains. Typical protein molecules are long spiral chains, suited to the fibrous structures they form.

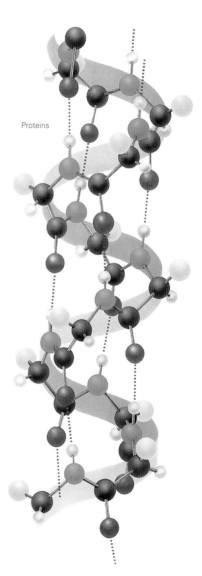

Proteins

Digestive System

FATS, OILS, AND LIPIDS

Fats are the second form of fuel for the metabolic furnace. They are usually regarded (particularly in the kitchen) as semi-solid but easily meltable substances, such as butter and lard; oils are seen in a similar way, but they are more liquid—although still viscous—at room temperature. To resolve any confusion, the preferred scientific name for the group is "lipids."

Just as proteins are made up of amino acids, and carbohydrates consist of glucose subunits, so fats are combinations of subunits. Fatty acids are basically long strings of carbon atoms with hydrogen attached. There is usually a chemical "handle" at one end of a fatty acid, which is an organic acid (similar to that in an amino acid) hooked up to a small molecule called glycerol, or glycerine. A single glycerol can take in three fatty acids, and for this reason dieticians speak of the types of fats found in our food as being made up of triglyceride subunits.

Lipids are for building and burning. Every cell in the body depends on the lipid content of its cell membrane to preserve the structural and functional integrity of this cellular "skin." Second, fat provides the most concentrated possible energy source within the diet. Its nutritional calorie content (a measure of energy obtainable per unit weight) is twice that of carbohydrate or protein. But too much fat is too much to handle: excess of certain types, particularly animal-derived fats, is implicated in the furring up of arteries and the risk of heart attack.

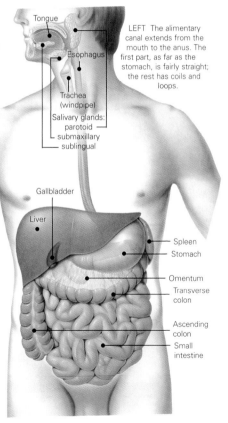

LEFT The alimentary canal extends from the mouth to the anus. The first part, as far as the stomach, is fairly straight; the rest has coils and loops.

Tongue
Esophagus
Trachea (windpipe)
Salivary glands:
  parotoid
  submaxillary
  sublingual
Gallbladder
Liver
Spleen
Stomach
Omentum
Transverse colon
Ascending colon
Small intestine

Digestive System

VITAMINS

Vitamins are a diverse group of chemicals, mostly smallish molecules, that serve as accessory ingredients in the many biochemical pathways of cellular metabolism. Their common feature is that, with a few exceptions, the body cannot manufacture them itself and so must take them ready-made from the diet.

MINERALS

Minerals are elements such as calcium, sodium, iron, iodine, and phosphorus. Some abound throughout the body, such as calcium, which is a major constituent of bones and teeth, and sodium, a vital functionary in the passage of electrical nerve impulses. Iron is essential for building the oxygen-carrying hemoglobin of red blood cells. Other minerals, the "trace elements," are needed in only small amounts. It is likely the body needs such tiny amounts, that some have yet to be discovered.

RIGHT Digestion—the chemical breakdown of foods into their component substances—begins in the mouth with the action of enzymes in the saliva and continues along the length of the digestive tract. Carbohydrates end up as glucose, proteins are broken down into their amino acids, and fats become fatty acids. Each is used to fuel metabolic processes, provide energy, or build new tissues. Excesses may be stored in the liver or tissues. Waste products are excreted via the kidneys.

THE LENGTH OF OUR INTESTINES
In adults, the intestines are at least 25 ft. (7.6 m) long, not as long as those of say, a fully grown horse, whose coiled up intestines measure nearer 89 ft. (27 m) long.

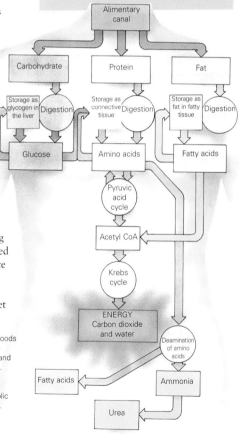

Digestive System

LIP SERVICE

Part your lips and you open the entrance to the digestive system. Beneath the skin of the flexible and fleshy lips is a circular muscle, the orbicularis oris, which is tagged on to the complex network of muscles in the face, jaw, and upper neck. Coordinated contractions of these muscles produce all lip and mouth movements, with their varied functions.
Some lip activities are communicative:

both visual, from the merest hint of a smile to an enormous grin; and auditory, from the pursed intake of breath that signifies doubt, to the wide-open funneling of a scream of fright or rage. Some are respiratory: from the slight crack of quiet breathing, to open-mouthed gasping for air after a short sprint. Some are sexual (kissing is one example). And some are digestive, from the delicate sucking of liquid through a straw, to ensuring that what has gone into the mouth, to be chewed into pulp, does not dribble out again.

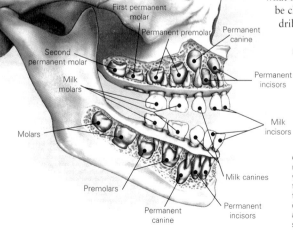

First permanent molar

Permanent premolar

Permanent canine

Second permanent molar

Milk molars

Permanent incisors

Molars

Milk incisors

Premolars

Milk canines

Permanent canine

Permanent incisors

LEFT A child's first, or milk teeth, (shown in diagram) are easily displaced by the permanent teeth (shown in blue) as they push through the gums. The eight adult premolars—two in each side of each jaw—replace the child's molars, while adult molars grow in previously unoccupied spaces at the back of the jaws. The dentition in the diagram includes eight adult molars. Four more, the so-called wisdom teeth, may erupt behind them. Incisors and canines, at the front of the jaw, are used for biting off and slicing food; the premolars and molars then crush it into pieces small enough to swallow.

Digestive System

A BITE TO EAT

The digestive journey commences in the mouth. It is the start of a long campaign to break down food. The advance troops are teeth, tongue, and saliva. Once in the stomach, the chemical infighting starts in earnest.

You may use knives, forks, and spoons to cut and trim your food into dainty, mouth-sized chunks. But your teeth are just as capable as cutlery. Each tooth is firmly embedded in a special outgrowth of the jaw bone known as the alveolar process, and is cushioned by gum (periodontal) tissue and fibrous connective matter. The part of the tooth within the jaw is the root; the section exposed above the gum line is the crown.

At the tooth's cove is its sensitive pulp, a tangled web of blood vessels that nourish the dental tissues, and nerve endings which warn of excessive pressure or pain. The pulp is shrouded in a layer of dentine, a fairly hard, bonelike material that makes up the bulk of the tooth's volume. Cloaking the dentine is enamel, the outermost whitish layer and the hardest substance made by the body. Odontoblast cells line the pulp cavity and manufacture the overlying dentine. Ameloblasts construct the enamel coating; they disappear once the enamel has been formed.

Like many of our mammalian cousins, we get through two sets of teeth in a lifetime (more if the dentist advises false ones). The dental primordia are laid

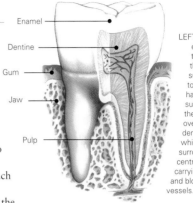

LEFT The enamel that forms the outer surface of a tooth is the hardest substance in the body. It overlays the dentine, which in turn surrounds the central pulp carrying nerves and blood vessels.

down during fetal life, and so poor diet or disease in the mother can have serious and even permanent effects on the teeth of her children. If she takes the antibiotic tetracycline, for example, her child may have discolored teeth.

The first, deciduous or "milk" teeth, 20 in number, erupt from about the seventh month after birth. The age range is very wide, however; a few babies already have growing teeth at birth, while others still have none at their first birthday. The second, or adult teeth, begin to erupt from the age of about seven years.

HOW LONG?
It takes from five to thirty seconds to chew food, then just a further ten seconds to swallow it.

Digestive System

DOWN THE SLIPPERY SLOPE

It is difficult to place a solid item in the mouth and then swallow it at once. Think of taking a medicinal tablet: the natural inclination is to chew, unless the tablet is "disguised" in water. Food in the mouth is masticated with saliva, a watery fluid rich in the starch-breaking enzyme amylase, which lubricates the processes of chewing and swallowing. Up to three pints of saliva are secreted daily. Three glands on each side of the mouth release copious quantities when stimulated by the smell and taste of food (or even the thought of a meal) and by its physical presence in the mouth. These glands are the parotid (over the angle of the jaw, near the ear), the submandibular (tucked into the lower jaw), and the almond-shaped sublingual (under the tongue).

IN YOUR STOMACH
Food can slosh around in your stomach for up to three to four hours before moving on to the intestines.

Gradually the food is chewed, pummeled and pulped, mixed with saliva into a moist mass called a bolus, and made ready for swallowing. At the start of a swallow, the tongue tip touches the

BELOW and OPPOSITE PAGE Chewing and mixing with saliva converts a mouthful of food into a soft ball called a bolus. Swallowing it involves a series of complex, but automatic movements of the tongue and throat muscles. First the tongue rises at the front against the roof of the mouth to push the bolus into the pharynx. The soft palate swings upward to block off the exit from the nose, and the epiglottis closes the entrance to the trachea, or windpipe, to prevent food from "going down the wrong way." Once the bolus is in the esophagus, peristalsis takes over and squeezes it along the length of the esophagus to the entrance to the stomach.

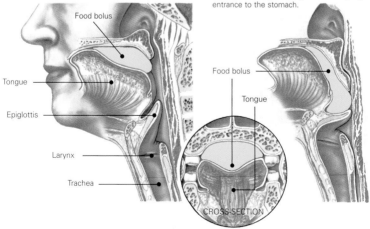

Food bolus

Tongue

Epiglottis

Larynx

Trachea

Food bolus

Tongue

CROSS-SECTION

Digestive System

roof of the mouth at the hard, front part of the palate—a shelflike plate dividing the nasal and oral cavities. Trapping the bolus behind it, the body of the tongue rises from the front to push the food back and down, into the pharynx (throat). The pharynx is an apparentus of muscles and specialized tissues used for accurate transport of the food to the esophagus (gullet)—most importantly past the entrance to the trachea (windpipe). A bolus entering the top of the trachea, rather than the esophagus, results in choking and possibly asphyxiation.

As the food reaches the pharynx it lifts the rear, soft part of the palate, which swings up like a one-way trapdoor to block its passage into the nasal cavity. Likewise, a flap on the lower front of the pharynx, the epiglottis, swings over and down to cover the entrance to the

trachea; simultaneously the larynx raises itself to narrow its entrance and hide it beneath the epiglottis. As all this occurs, further muscular activity creates propulsive waves in the pharynx, massaging the bolus down into the esophagus.

The whole process of swallowing is controlled by a series of reflexes that ensure food and liquids pass smoothly into the next region of the digestive tract. We swallow hundreds of times each day, hardly ever bothering to think about it, yet each event is a potential disaster if food goes down "the wrong way."

IN A LIFETIME
In your lifetime, your digestive system will probably have to handle as much as 50 tons of food.

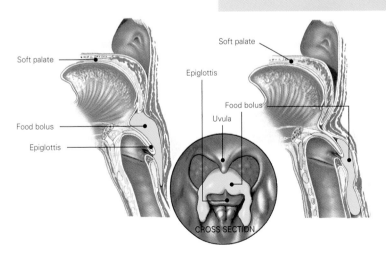

Soft palate

Soft palate

Epiglottis

Food bolus

Food bolus

Uvula

Epiglottis

CROSS SECTION

Digestive System

THE FATE OF THE BOLUS

The whole of the gut, from the top of the esophagus to the rectum, can make snakelike writhing movements whereby the digestive contents are progressively pushed through the system. The principal components of this action, known as peristalsis, are the same throughout the gut, though their speed and force vary from one section to another.

Peristalsis is brought about by layers of muscles and nerves within the gut wall. In the outer layer, the muscle fibers are arranged lengthwise, pointing along the gut tube. The inner layer consists of fibers encircling the gut. Between the two layers is the myenteric plexus, a network of delicate nerve fibers. This is connected to another plexus, the submucus plexus, just next to the innermost mucous lining of the gut.

A lump of food or a bubble of gas in the central space—the lumen—of the gut stretches its wall. This initiates

RIGHT Peristalsis is a wave of involuntary muscular contractions that forces material along the digestive tract. Food is moved along the esophagus, and partly digested food and feces are squeezed along the intestines.

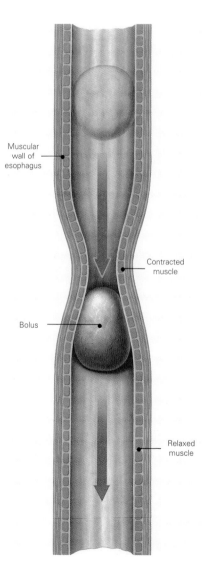

Muscular wall of esophagus

Contracted muscle

Bolus

Relaxed muscle

MOVING ON TO THE LARGE INTESTINE
After food has gone through the small intestine, it travels to the large intestine where it can hang out and dry up from anything from eight hours to two days.

Digestive System

contraction of the muscle immediately behind the lump, which eases the lump forward into the next segment of the plexus, where the muscles are relaxed. Here the process repeats as this next segment of wall stretches. In this way waves of contraction flow along the gut, followed by waves of relaxation. Peristalsis in the pharynx and esophagus is so powerful that if necessary you can swallow food into your stomach even when upside-down. Reverse peristalsis can be initiated in times of trouble, causing the stomach contents to travel up the esophagus and be thrown out of the mouth with surprising force. We call it vomiting, and it is intended to rid the digestive system of bad foods, poisons, or excessive contents.

SHUTTING OFF THE SYSTEM

Peristaltic waves travel down the 10 in. (25.4 cm) of esophagus at about 2 or 3 in. (5 or 7.6 cm) per second. At the base of the esophagus is a ring of muscle, the esophageal sphincter. Like peristalsis, the actions of sphincters are useful features of the digestive tract. These rings of specialized muscle are capable of sustained contraction to close an orifice and seal off one section of gut from the next, while the contents are processed. Under automatic nervous control, the muscles relax occasionally and allow the contents to ooze through.

The esophageal sphincter is particularly important because the next region of the tract is the stomach—and this is filled with powerful, churning acid. If the sphincter weakens, or if there is high intra-abdominal pressure (as when bending forward or in obese people), stomach contents may well up into the lower part of the esophagus. The esophageal wall is not nearly so resistant as the stomach is to acidic contents, and the welling-up produces an unpleasant, bitter sensation deep in the throat. This is known as heartburn—although it has nothing to do with the heart.

PEANUT BUTTER
Americans eat about 700 million pounds of peanut butter in a year.

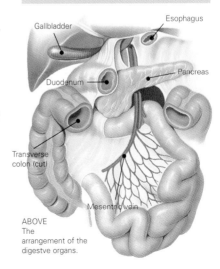

ABOVE
The arrangement of the digestve organs.

Gallbladder

Esophagus

Pancreas

Duodenum

Transverse colon (cut)

Mesentric vein

Digestive System

THE ACID BATH

The average adult stomach holds from 1.14 l to 3.41 l (2 to 3 pints) and manufactures the same volume of "gastric juices" every 24 hours. The stomach plays several roles in digestion.

It is a food hopper, or storage reservoir. The upper baglike portion, the fundus, holds a hurriedly-eaten meal and feeds it part at a time to the lower portion, the antrum.

It is a food-mixer. Strong muscles, principally in the antral wall, contract to squash and pulverize the contents into a sticky slushy mass called chyme.

It is a sterilizing unit. Parietal cells, in glands in the stomach lining, make powerful hydrochloric acid that kills many of the germs in unwisely consumed, contaminated food.

It is a digesting tub. The acid, along with the protein-splitting enzyme pepsin made by other cells in the stomach wall, sets to work to split and crack the chemicals in food.

And it is well protected from itself. A thick layer of mucus, secreted by yet other cells in its lining, coats the inside of the stomach and stops it digesting itself. If there is too much acid, or the mucous coat is deficient, the acidic contents erode raw spots in the stomach wall. These are gastric ulcers.

RIGHT A vertical slice through the stomach and intestines shows how they are supported by the membranes of the mesentery and omentum.

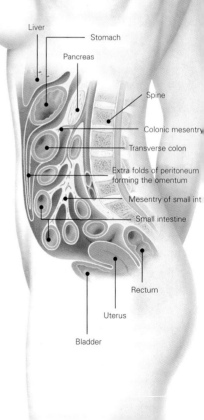

Liver

Stomach

Pancreas

Spine

Colonic mesentry

Transverse colon

Extra folds of peritoneum forming the omentum

Mesentery of small int

Small intestine

Rectum

Uterus

Bladder

Digestive System

GETTING THE JUICES FLOWING

The control of acid and enzyme secretion in the stomach is crucial, because it must be coordinated with the appearance of the food on which these substances work. The organ is well supplied with nerves, most of which come from the autonomic nervous system—that part of the overall nervous system that controls automatic or involuntary movements, mostly involving the internal organs. The most important

nerve is the vagus, which runs from the brainstem or medulla, at the base of the brain, down through the chest and abdomen to control various automatic functions. The vagus carries nerve impulses directly from the brain to the groups of cells in the stomach lining which secrete gastric juices.

THE VILLI

Like sponges, villi absorb tremendous amounts of nutrients from the food you eat, then these nutrients flow into your bloodstream.

BELOW Tiny finger like projections called villi, each served by a tiny artery and a vein, whose capillary junctions enfold a lacteal tube which carries away the products of fat digestion.

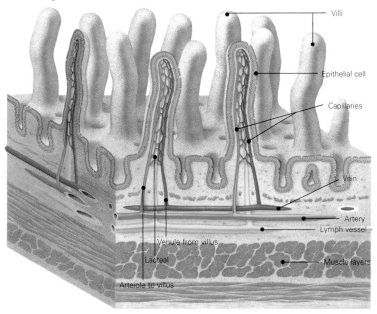

- Villi
- Epithelial cell
- Capillaries
- Vein
- Artery
- Lymph vessel
- Muscle layers
- Venule from villus
- Lacteal
- Arteiole to villus

Digestive System

The smell and taste of food, and its physical presence in the mouth, esophagus, and stomach, all help to get the juices flowing.

Hormones play their part, too. Nervous stimulation of cells in the upper part of the stomach wall causes the cells to release the hormone gastrin, which spreads through the blood supply and encourages further release of juices into the stomach cavity. Therefore, soon after the chewed food enters the stomach, it is assailed by various chemicals designed to break it down.

So much for filling. What of emptying? The same dual nervous-hormonal control is in evidence. The pyloric sphincter at the stomach's exit relaxes occasionally to let squirts of chyme through into the duodenum, the first 10 in. (25.4 cm) section of the small intestine. In the enterogastric reflex, the squirts of chyme gradually fill the initial part of the duodenum and distend its wall. Sensors in the wall initiate the reflex, sending nerve signals along the

vagus, and other small nerves of the celiac plexus network, back to the stomach. The signals dampen the stomach's movements so that the amount of chyme flowing from it decreases.

As chyme slops into the duodenum, it also triggers the release of the hormones secretin and cholecystokinin from the mucosal lining there. These hormones assist in limiting gastric motility and acid secretion, as well as stimulating organs such as the gallbladder and pancreas to prepare for the coming meal.

THE TRAP DOOR
Being the flap at the back of the tongue, the epiglottis stops food going down the windpipe to the lungs. When you swallow, the epiglottis automatically closes, and when you breathe, the epiglottis opens so air can go in and out of the windpipe.

BELOW For efficient absorption of the products of digestion, the internal surface of the intestines is designed to have maximum area. There are three successive levels of maximization. First, the mucous membrane lining is gathered up into epithelial folds. Second, each fold bears thousands of villi. Third, the individual absorptive cells bristle with even smaller microvilli.

Epithelial fold

Villi

Microvilli

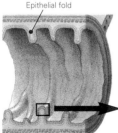

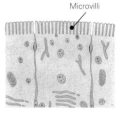

Digestive System

CHEMICAL BOMBARDMENT IN THE INTESTINE

The small intestine has three main sections. First is the duodenum, already mentioned. Second is the jejunum, some 8 ft. (2.5 m) long and coiled behind the area of the navel. Third is the ileum, 3 ft. (0.9 m) longer, which loops its way down to the lower right of the abdomen. Here it swells to form the cecum, the first part of the large intestine. Each section of small intestine is about 1 in. (2.5 cm) in internal diameter.

The initial attack by acid in the stomach does a great deal to break down food, but most "digestion" (in the chemical sense) and absorption of nutrient molecules takes place in the small intestine. The liquefied food that is chyme flows through the duodenum and there mixes with a battery of enzymes pouring

GALLSTONES

Gallstones are a common disorder of middle-aged people, and if the gallbladder becomes inflamed it causes the pains of cholycystitis. Surgery to remove gallstones is now routine, and if necessary the whole gallbladder can be removed.

in along a duct from the pancreas. Now the food is under heavy bombardment. Among the bigger guns are the pancreatic enzymes trypsin and chymotrypsin, which pound proteins into shorter and shorter

BELOW Bacteria and toxic substances in the intestines are dealt with by M, or microfold, cells in the surface epithelium of the villi. These defensive cells direct invaders—recognized as "foreign" proteins—toward a neighboring lymphocyte, which produces antibodies to neutralize the invaders.

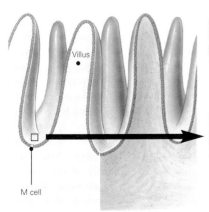

Villus

M cell

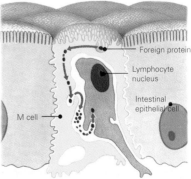

Foreign protein

Lymphocyte nucleus

Intestinal epithelial cell

M cell

Digestive System

segments until they are only a few amino acids long. Pancreatic amylase, another powerful enzyme, sets about the starches and other large carbohydrate molecules, and breaks them into smaller sugar molecules. The pancreatic juices also contain alkalis, which neutralize the corrosive acid of the stomach and allow these various enzymes to work at their peak efficiency in the now slightly alkaline environment.

Dietary fats present more of a problem. Another component of pancreatic juice, the lipid-splitting enzyme lipase, is ready to enter the fray. But its target is awkward. Because fats are not water-soluble, they tend to split up and regroup in the intestine as small globules, which lipase finds difficult to penetrate.

The globules must be broken into much smaller droplets, a process chemists call emulsification. Enter bile, literally, along the bile duct leading from the gallbladder. Bile is an especially effective emulsifying agent and splinters fat droplets, so that the fragments can be more easily picked off by lipase.

PERISTALSIS
Being involuntary, this is the rythmic muscle movement which forces food in the esophagus from the throat into the stomach. Peristalsis also allows you to eat and drink while upside-down.

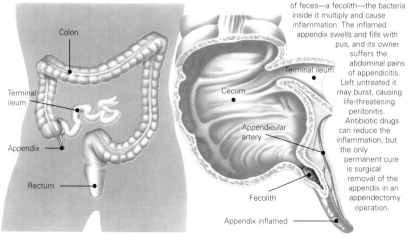

LEFT, BELOW and RIGHT A useless nuisance, the appendix is a blind-ended tube off the lower part of the cecum. If it becomes blocked by a hardened piece of feces—a fecolith—the bacteria inside it multiply and cause inflammation. The inflamed appendix swells and fills with pus, and its owner suffers the abdominal pains of appendicitis. Left untreated it may burst, causing life-threatening peritonitis. Antibiotic drugs can reduce the inflammation, but the only permanent cure is surgical removal of the appendix in an appendectomy operation.

Colon

Terminal ileum

Appendix

Rectum

Terminal ileum

Cecum

Appendicular artery

Fecolith

Appendix inflamed

Digestive System

THE MESENTERIC CONNECTION

The small intestine does not float freely inside the abdomen. If it did, there would be every chance that it could twist and knot, blocking the passage of food. Instead, its multitudinous loops are anchored to the rear wall of the abdomen by a tissue structure known as the mesentery. The appearance of this thin, membraneous sheet has been likened to an open fan, rooted by its "handle" to the back wall of the abdomen, and with its long, free edge supporting a great length of intestine. The mesentery not only holds the intestine in place, but also carries to it a blood supply (since, like any other organ, the intestine needs a supply of oxygen and suitable nutrients). In addition, the mesentery supports a vast

TWO TYPES OF NERVE CONTROL
There are two types of nerves to help control the action of the digestive system; the extrinsic (outside) nerves, and the intrinsic (inside) nerves.

network of blood and lymph vessels which transport the absorbed nutrients away from the intestine.

The abdominal cavity is lined by a thin membrane, the periosteum. This folds inside itself and extends to cover each digestive organ, in an almost impossibly complicated series of curves, folds, and invaginations.

The battle is won, but the war is not over. Food popped into the mouth perhaps eight or ten hours previously is, at last, truly "inside" the body, broken into its smallest viable subunits and floating along in the bloodstream and lymphatic channels. Yet the story continues. The food molecules, now "prisoners of war," are shipped to the next port of call: the liver.

Terminal ileum

Cecum

Fecolith

Appendix inflamed and swollen

Terminal ileum

Cecum

Fecolith

Perforated burst appendix

Digestive System

THE ULTIMATE FOOD PROCESSOR

The liver is the body's central "food processor." Chemically it is extremely active, playing host to more than 500 metabolic pathways—and doubtless there are still many more to be discovered. Some pathways involve the breakdown of complex chemicals (catabolism); others involve synthesis (anabolism), especially of protein molecules. The liver acts as a cleansing station for the blood, inactivating hormones and drugs. The Kuppfer cells that line the liver's blood vessels mop up unwanted elements and infectious organisms reaching it from the gut. In particular, the liver has three principal functions.

There is a key role in carbohydrate metabolism. Glucose from the intestines is chemically condensed to form glycogen ("animal starch"), which is laid down in the liver itself and in other organs. Excess glucose is converted and shunted off to adipose cells around the body, to be stored as fat.

There is a key role in fat metabolism.

The lipid products of digestion arrive in the liver and certain of them are used to manufacture various vital fatty substances, notably cholesterol. There is an essential ingredient in the construction of some hormones and in nerve cell functioning, although it has become a villain in the story of atheroma and heart disease.

Finally there is a key role in protein metabolism. Unwanted proteins are disassembled into their amino acid subunits, which are broken down still further and rebuilt into other amino acids as the current requirements dictate. These amino acids are then reassembled in the correct sequence to make new, wanted proteins.

Proteins are distinguished from carbohydrates and fats by their nitrogen content (in addition to carbon, hydrogen, and oxygen). While it is rearranging amino acids and proteins, the liver produces a certain amount of "free" nitrogen, surplus to requirements. This is rapidly converted into urea—the main waste product from the liver's chemical juggling of proteins—which then enters the bloodstream, to be filtered out by the kidneys and excreted in the urine.

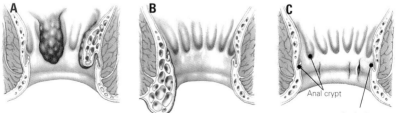

A   B   C

Anal crypt

Anal valves

Digestive System

JOURNEY'S END

The final region of the gut is the 2 in. (5 cm) long anal canal. Of all the sphincter mechanisms in the digestive tract, this is probably the best known. So are its possible disorders, such as hemorrhoids or "piles," fissure, fistula, and abscess.

In fact there are two anal sphincters, internal and external. The internal sphincter is normally in a contracted state to prevent any leakage of fecal material through the anus. As the rectum becomes stimulated by the accumulation of feces, nervous impulses pass from it to the anal area and trigger relaxation. Of course, in an adult this is rarely a simple reflex reaction. Rectal distension also produces a conscious desire to empty the bowels. The external anal sphincter has a large degree of voluntary nervous control, so that we can keep this contracted until we will it to relax, allowing defecation to take place.

There is no "normal frequency" of defecation, although each individual tends to have his or her own pattern, and a sustained departure from this pattern might provoke suspicion. Perhaps the old saying, "from once every three days to three times a day," represents something like the range across which a physician would not show concern. In some people, worry or anxiety can heighten the sensitivity of the system, necessitating frequent visits to the bathroom.

DIFFERENT SYSTEMS
Carnivores have the simplest digestive system in mammals, humans have a much more extensive large intestine. Cattle and sheep have large sets of forestomachs before reaching the stomach.

LEFT Internal hemorrhoids (A) occur near the entrance of the anal canal. External hemorrhoids (B) are accompanied by swelling and cause considerable pain during defecation. The consequent reluctance to defecate can, in turn, lead to constipation, which only makes the condition worse. An anal fissure (C) involves splitting of the skin that lines the anal canal, again resulting in bleeding and painful defecation. Severe hemorrhoids may not need treating with surgery; and fissures often heal spontaneously.

RIGHT The contents of the rectum are confined by the internal and external sphincters of the anus. The area is well served by veins, and damage to them causes the bleeding of hemorrhoids.

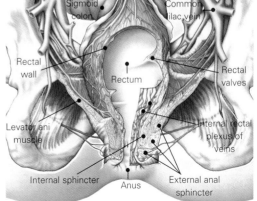

Sigmoid colon

Common iliac vein

Rectal wall

Rectum

Rectal valves

Levator ani muscle

Internal rectal plexus of veins

Internal sphincter

Anus

External anal sphincter

ENDOCRINE SYSTEM

The human body is a complex system of interrelated organs and tissues, all of which must work together if it is to function properly. A wide variety of physiological processes—ranging from growth and development to digestion and reproduction—have to be continuously monitored and controlled. Both monitoring and control have to take place without conscious intervention, and these are achieved by the endocrine system, which carries them out with the help of a highly ingenious array of chemical messengers called hormones.

Released into the bloodstream, the body's liquid communication system, hormones are carried to all parts of the body where they trigger the required action. But the effect of any particular hormone must be carefully controlled—if it goes on working too long or too vigorously it could disrupt the delicate balance of the body. The secret of this control mechanism is feedback, which is used in much the same way as it is in machines—the output of the process is sensed and fed back to modify the action of the initiating agent.

Many processes use negative feedback; the sensed output is fed back in such a way as to reduce or switch off the input, and as a result the system settles down to a stable condition. But if the output is disturbed by some external influence, this reacts on the input and the system responds so as to restore the stable state. A common example is a thermostat that controls a heating system by turning the heat on and off to maintain a particular temperature. Less common is positive feedback, in which the sensed output adds to the input, causing even more output that goes on growing until the limits of the system are reached.

Within the endocrine system negative feedback is normal. The organ responsible for outputting a hormone senses the level of hormone in the bloodstream, or the level of the substance being controlled by the hormone. When the level rises too high, the hormone output is reduced or switched off, or a counteracting hormone released to bring the level down. Conversely, if the level is too low the hormone output is stepped up or switched on. This delicate balancing trick ensures that the body always has the correct hormone levels it needs.

HORMONE FACTORIES

Most hormones are proteins and they are manufactured in endocrine glands located in various parts of the body. The major ones include the pituitary (at the base of

THE ENDOCRINE SYSTEM USES CHEMICALS
While your nervous system uses electricity to orchestrate all sorts of things in your body, your endocrine system does even more through the wonder of chemicals.

Endocrine System

the brain), thyroid and parathyroids (in the neck), pancreas and adrenal glands (in the abdomen), and the gonads, or sex organs. The main feature of the glands is that the hormones they produce are secreted into the bloodstream (or sometimes other tissue fluids), in which they are carried through the body to act on various organs and tissues. Other glands in the body release their secretions into tubes or ducts which carry them to the site of action. These are the exocrine glands. Some glands—for example, the pancreas—work in both ways and are known as mixed glands.

HORMONES IN ACTION

When a circulating hormone reaches its target cell it keys in, unlocks the system, and then switches on the appropriate cell action. This switching process can take a number of different forms. In one it affects the membrane of the target cell so that it allows specific ions or molecules to pass through and start up the cell processes. In another the hormone binds on to receptors on the cell membrane and triggers the release of a secondary messenger that sets the cell to work. Steroid hormones can pass right

through the cell wall and react directly with receptors in the cell cytoplasm.

After they have triggered the required cell action the hormones are of no further use and have to be cleared out of the system. Commonly they are transported to the liver to be broken down, or may be destroyed by the target tissues themselves. The water-soluble compounds that remain are efficiently disposed of by the body's excretory system.

ENDOCRINE GLANDS
Endocrine glands manufacture chemicals called hormones that are released directly into the bloodstream, traveling to all parts of the body and acting as chemical messengers.

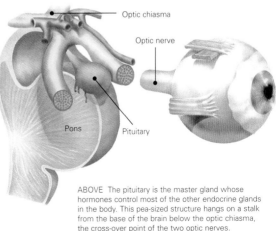

Optic chiasma

Optic nerve

Pons

Pituitary

ABOVE The pituitary is the master gland whose hormones control most of the other endocrine glands in the body. This pea-sized structure hangs on a stalk from the base of the brain below the optic chiasma, the cross-over point of the two optic nerves.

Endocrine System

THE MASTER GLAND

The most important endocrine gland is the pituitary. It is the master gland, and has a controlling effect on many of its colleague glands. About the size of a pea, it is suspended from the brain by a stalk so it sits just above the roof of the mouth, more or less in line with the bridge of the nose.

The thyroid produces hormones that control metabolism, the intricate biochemistry involved in the breakdown of food into useful substances and energy, and the disposal of waste products of the body. It is a fleshy gland positioned in the neck just below the larynx. Four parathyroid glands are attached to the back of the thyroid. They control the body's levels of calcium and phosphorus, chemicals needed for healthy bones and teeth.

Located on the upper part of each kidney, the two adrenal glands are roughly triangular in shape. Their hormones control fluid and mineral balances in the body, the metabolism of glucose, and one of the body's most fundamental life-preserving functions—its response to stress and danger.

The pancreas is situated in an inaccessible position at the back of the abdomen, underneath the liver and in front of the vertebral column. It is mainly involved in the production of digestive enzymes, which pass along ducts to the duodenum at the exit of the stomach. But the pancreas also produces the important hormone insulin, which controls the level of the energy-generating sugar glucose in the blood.

The gonads have different functions in men and women, though there are two in each sex. Men have external gonads, the testes, or testicles. Women have ovaries, almond-shaped organs about one and three-quarters of an inch long located within the pelvic girdle. Sex hormones—estrogens, progesterone, testosterone, and androsterone—are produced by both males and females. But the end results

RIGHT The thyroid gland is located in the front part of the neck, partly surrounding the larynx, with the four parathyroid glands embedded at its rear. The thyroid hormones control metabolism.

Thyroid cartilage

Thyroid

Carotid artery

Vagus nerve

Trachea

Jugular vein

HORMONES

"Hormone" means to "excite" or "spur on" and that is exactly what hormones do, they cause things to start happening.

Endocrine System

of the actions of these hormones are very different in the two sexes because each produces significantly different proportions of each hormone. Predominance of testosterone and androsterone defines maleness, whereas estrogen and progesterone are the predominants in females.

Bodily health and balance depend vitally on the proper operation of the endocrine glands. Faulty behavior of just one of them can have profound effects on the whole body. In growing children malfunction of the pituitary, for example, may cause them to end up as dwarfs, or as giants. Similarly incorrect levels of the sex hormones result in inadequate or misdirected development. This may lead to infertility or the production of incorrect secondary sexual characteristics—such as the growth of facial hair in women or of breasts in young men.

Tumors can affect the endocrine glands, in the same way as they sometimes attack the rest of the body. But even non-cancerous tumors are liable to cause malfunction of the endocrine system and result in overproduction or underproduction of hormones. In such cases surgery or radiotherapy may be needed to deal with the tumor. Luckily it is also possible to treat many endocrine disorders by a simple process of hormone replacement therapy. The missing hormone is supplied in tablet or injectable form to reestablish the correct levels in the bloodstream. The best-known example of such therapy is the regular administration of the hormone insulin to people with diabetes mellitus, caused by lack of that hormone.

LOTS OF HORMONES
You have over 30 hormones busily orchestrating and regulating different activities in your body.

CHEMICAL MESSENGERS

Using the bloodstream to carry messages between various organs is a comparatively slow but generally certain process, which can work locally or at a considerable distance from the place where the messages originated. But because all of the hormones circulating in the blood get carried to all the organs in the body, it is important that each of them has a very specific action. If the same hormone had two or more different functions, they would be bound to interfere with one another and it would be impossible for the vital regulatory feedback mechanisms to work.

To make sure that they do the correct jobs, the hormones act like keys. Each hormone is recognized by and fits, or unlocks, a specific target cell on the organ or tissue it is supposed to activate. The hormone can only affect the cells that recognize it, and the cells can only be affected by their specific hormones.

Chemical building-blocks are molded together to make the individual

Endocrine System

hormones, which come in three main types. Most common are the protein and peptide hormones formed from amino acids. The substance cholesterol—abundant in animal fats and dairy products—is the basis of the steroid hormones. And there is a third group of miscellaneous hormones which do not really fit into either of the other two groups.

The actual production of hormones in the body is a simple manufacturing process. To begin with, blueprints for the protein and peptide hormones are produced in the nucleus of the relevant endocrine cell. From the "drawing office" the plans, in the form of ribonucleic acid (RNA), are passed out to an assembly plant—ribosomes in the body of the cell. In this assembly operation, known as transcription, the RNA controls the build-up of the protein and peptide molecules that make the hormones. Finally the hormone molecule is packaged into a pouchlike vesicle by the cell Golgi apparatus.

When needed, the molecules are expelled through the cell wall and

WHAT HORMONES REGULATE
Just a few of the things your hormones regulate include: your body temperature, how you sleep, and whether you are fat or slim.

BELOW A chain of hormone reactions can be triggered by the hypothalamus in the brain, and controlled by a feedback mechanism. Initially (1) a hormone from the hypothalamus causes the pituitary to release one of its hormones (red arrows, 2). The pituitary hormone travels in the bloodstream to a target gland somewhere in the body, stimulating it to release a third hormone (blue arrows, 3), which is also carried in the blood to where it is needed. Some of it reaches the hypothalamus and "switches off" its pituitary-stimulating action (4).

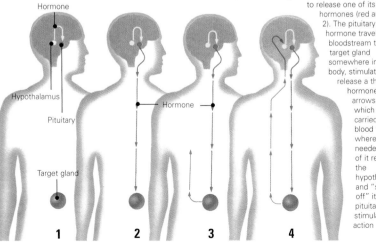

Hormone

Hypothalamus

Pituitary

Hormone

Target gland

1   **2**   **3**   **4**

Endocrine System

effectively thrown into the bloodstream. Some hormones are made in a simpler way from "dummy" or precursor molecules. These molecules have no function of their own but can be stored in the cell and quickly transformed to give their corresponding hormone molecule.

Like burgers in a fast-food restaurant, steroid and miscellaneous hormones are made to order. When the body needs them, an enzyme in the cell is activated and sets off a series of reactions to build the hormone molecule from building blocks stored in the cell cytoplasm. Once synthesized, the steroid molecules pass easily through the cell membrane and straight into the bloodstream for distribution.

Not all of the hormones circulating in the blood are needed for immediate action. Some are banked to form a hormone savings account that can be

drawn on as required. The banking works by binding the hormone to a protein so it is not directly active but can continue to circulate in the bloodstream. When the hormone is needed the binding is easily broken and the hormone released for use. Another of the system's devices is to make some of the hormones in one form and then process them by another body organ to make a different hormone with new functions.

PUBERTY
When your body begins its final journey to become an adult—puberty, this is also managed by hormones. This involves all sorts of small and big changes, in your brain as well as your body.

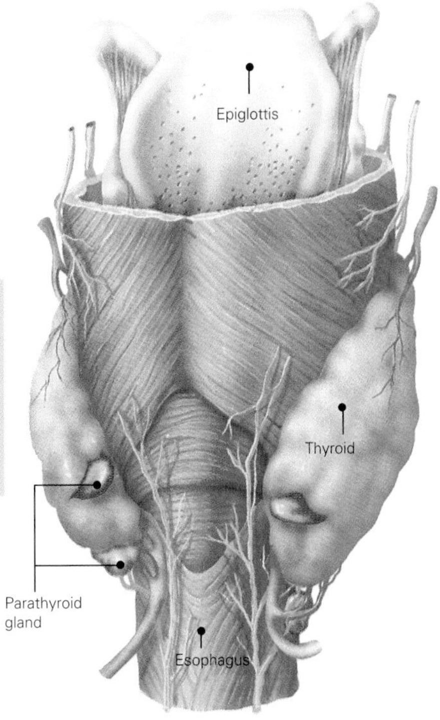

RIGHT The parathyroid gland. The parathyroid hormone maintains the calcium levels in the blood.

Epiglottis

Thyroid

Parathyroid gland

Esophagus

Endocrine System

PITUITARY IN CHARGE

Overall control of the workings of the endocrine system is the responsibility of the pituitary gland. This conductor of the hormonal orchestra works in close cooperation with the hypothalamus, an organ which, like the orchestra leader, provides an all-important link between the brain, nervous system, and endocrine system. The pituitary exercises its unique control by producing hormones that act on the other endocrine glands and stimulate them to release their own hormones.

Control of the individual glands involves both increasing and reducing their output, and many of these processes are regulated by a feedback loop involving the hypothalamus. Sensor cells in the brain, like the thermostat in a room heating system, monitor the levels of the various hormones in the blood and signal the results to the hypothalamus, which then modifies the activity of the pituitary. In turn this affects the action of the various glands, which reduce or increase their output so as to maintain the required balance.

The essential communication between hypothalamus and pituitary is also achieved by the use of chemical messengers, or hypothalamic hormones, produced by neurosecretory cells (nerve cells which make hormones) in the hypothalamus. They are released into the local bloodstream and pass to the pituitary, where they control its activity. Some of the hormones also pass into the general circulation, while similar products are manufactured in other areas of the brain and other organs of the body. The complex interactions of these hypothalamic hormones and the ways in which they help regulate the body's systems is even now not fully understood.

About the size of a pea, the pituitary is suspended from the brain by a slender stalk and sits in a bony pocket just above the back of the nose. Even though it is a small organ it is divided into two main parts, the anterior (front) and posterior (rear) lobes, which work in very different ways.

DIFFICULT TIME

Puberty can be a difficult time for some, including mood swings, as your body has to adjust to all those molecules and hormones racing around the body.

Bile duct

Pancreas

Duodenum

Jejunum

LEFT The pancreas lies just below the liver. Insulin is produced in the pancreas, which controls the way the body uses the blood sugar glucose.

Endocrine System

In a woman who has just had a baby, breast-feeding makes the anterior part of the pituitary gland produce prolactin, the second pituitary hormone with a direct action. Prolactin stimulates the mother's breasts to produce milk, thus the action of the baby goes on ensuring that she has milk to provide for her infant. The prolactin has a restricting effect on the actions of the female sex hormones and so tends to reduce the woman's fertility—women who are breast-feeding rarely have menstrual periods. The reduced risk of pregnancy is nature's way of minimizing the demands made on the feeding mother.

Oxytocin from the posterior portion of the pituitary is also released by the stimulation of breast-feeding. This hormone makes the milk ducts in the breast contract so the milk literally squirts out of the nipple into the baby's mouth. A new mother trying to breast-feed a baby may find it very difficult to get her milk flowing. Worrying about her failure creates stress and, paradoxically, can inhibit the action of the oxytocin, so that milk does not flow as it should. Relaxation is the key—a calm approach lets the oxytocin get to work as it should and the baby gets its feed.

The sex glands are stimulated by the gonadotropins, two of the other hormones produced by the pituitary. These are luteinizing hormone (LH) and follicle-stimulating hormone (FSH).

The remaining two pituitary hormones are corticotropin (or adrenocorticotropin,

ACTH) and thyrotropin (thyroid-stimulating hormone, TSH). As its name suggests, thyrotropin acts on the thyroid gland. Corticotropin influences the adrenal glands to make them increase the secretion of glucocorticoids. In turn, the glucocorticoids act on the pituitary to inhibit further production of corticotropin, giving a direct feedback loop that maintains the required glucocorticoid level.

BOYS AND GIRLS
For girls, puberty usually starts between the age of 10–15, whereas a boy will generally start puberty between 14–18.

BELOW The two triangular adrenal glands sit at the top of each kidney. Like all endocrine glands, their hormones are carried around the body in the bloodstream and for this reason they have a plentiful supply of blood through direct connections with the main circulation. The principal adrenal hormone, epinephrine (adrenaline), prepares the body to meet stress or threat.

Right adrenal gland

Superior mesenteric artery

Right kidney

Aorta

Endocrine System

Impotence and loss of sex drive in men is sometimes caused by the production of excess prolactin in the pituitary. The same fault causes infertility in women and may make their breasts start making milk even if there is no baby to be fed. These are typical examples of the complications that can arise when the operation of the pituitary is disturbed. Tumors are one of the more common causes of such disturbance, which often results in excess hormone production.

As well as acting directly, the excess may affect the operation of the other endocrine glands. An example is the excessive production of growth hormone, which causes giantism in children and acromegaly in adults. The usual treatment is to cut out the tumor by surgery or destroy it with radiotherapy. If surgery is undertaken the surgeon generally gets at the pituitary by working through the patient's nose. However, treatment of tumors may also damage the rest of the

gland, and when this occurs, the patient may need continued hormone replacement therapy for life to compensate for the damage.

The posterior lobe of the pituitary is much smaller than the anterior lobe and its function is to store and release the hormones oxytocin and vasopressin, which are produced by the neighboring hypothalamus. As well as being vital to the release of milk from the breast of a feeding mother, oxytocin also stimulates the contraction of smooth muscle, most especially that of the uterus. Indeed it is one of the agents that causes contractions when a baby is being born and can in fact

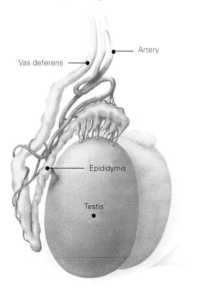

RIGHT Male sex hormones are made in the testes, a pair of egg-shaped glands located in the scrotum. The chief hormone, testosterone, brings about the development of secondary sexual characteristics in boys after puberty, and thereafter maintains a man's sex drive and maleness. The testes also produce and store sperm.

DIFFERENT HORMONES
Hormones are grouped into three classes based on their structure:
1. steroids, 2. peptides, 3. amones.

Artery

Vas deferens

Epididymis

Testis

Endocrine System

be used to induce labor at the end of a normal pregnancy. Oxytocin released when a newborn baby suckles on its mother's breast also causes contractions of the muscles of the womb and can help with ejection of the placenta after birth. Releases of more oxytocin with every breast-feed continues this muscular action and helps the womb return to its normal size.

Vasopressin is also known as an antidiuretic hormone and is involved with the control of blood volume and salt concentration. If blood is lost, or there is an increase in the salt concentration, the condition is detected by receptors in the brain which trigger the release of

vasopressin. The hormone then acts on the kidneys so that they absorb more water back into the blood and produce a more concentrated urine. This helps to dilute the blood salts and to increase the blood volume, so restoring the body to a healthy, well-balanced condition.

STEROIDS

Steroids are lipids derived from cholesterol. While testosterone is the male sex hormone, estradiol, similar in structure to testosterone, is responsible for many female sex characteristics. Steroid hormones are secreted by the gonads, adrenal cortex, and placenta.

Fallopian tube

Uterus

Ovarian vessels

Ovary

Fimbriae

ABOVE Female sex hormones are made in the ovaries, a pair of glands located in the abdomen on each side of the uterus. Initial production of these hormones, mainly estrogen and progesterone, brings about the development of secondary sexual characteristics in girls after puberty. During a woman's childbearing years they are also responsible for controlling the menstrual cycle, the approximately 28-day sequence of subtle bodily changes that includes, at about its midpoint, the release of an egg (ovum) from one of the ovaries. This event, called ovulation, can be prevented if a women takes hormones in the form of the contraceptive pill or injection; it ceases during pregnancy and at the menopause.

Endocrine System

THYROID GLAND

The rate at which the body's cells work—their metabolism—is controlled by hormones produced in the thyroid gland. The hormones concerned are thyroxine (T_4) and triiodothyronine (T_3), and most of them are bound to proteins in the bloodstream. The small amounts of the hormones that remain free act to influence the rate of activity of the body cells, with the protein-bound hormones acting as a reserve supply which is released as the free hormones get used up. Although the exact action is not certain it is believed that the cell action is actually triggered by T_3, and T_4 is converted to T_3 in other parts of the body to act as an additional reservoir.

Production of both these hormones in the thyroid is stimulated by the hormone throtropin (TSH) from the pituitary gland. Control of the process is another feedback operation carried out by the hypothalamus. In response to increasing thyroid hormone levels it acts on the anterior lobe of the pituitary, which in turn reduces the TSH level. This then cuts back the level of stimulation of the

LIFE WITHOUT THE ENDOCRINE SYSTEM
Imagine your body without glands. You'd be without all the ooze, sweat, mucus, chemicals, and juices in your body which make your body home.

thyroid, and so diminishes the amounts of hormones produced. Calcitonin is a third hormone produced in the thyroid and it acts to help control the calcium levels in the body.

The key chemical element in the thyroid hormones is iodine. If there is a shortage of iodine in the diet, the thyroid gland becomes more and more enlarged in its attempts to gather sufficient iodine to maintain normal levels of the hormones. Such enlargement creates a characteristic swelling of the neck known as a goiter. To reduce the likelihood of this happening, iodine compounds are often added to table salt, to make sure there is a sufficient supply in the diet. Iodine deficiency is not, however, the only cause of goiters. Thyroid disease is the second most common endocrine disorder after diabetes. Excess activity of the gland, with the production of too much of the hormones, results in hyperthyroidism or thyrotoxicosis. Graves' disease is a specific type of this condition which is characterized by swelling of the thyroid—leading to the development of a goiter in severe cases—and possibly bulging eyes that give the patient a typical staring expression. The high levels of thyroid hormones in the blood stimulate the body cells into increased action, resulting in a higher metabolic rate than normal. This causes excessive sweating, loss of weight, and hunger. There may also be heart problems and emotional upsets.

Endocrine System

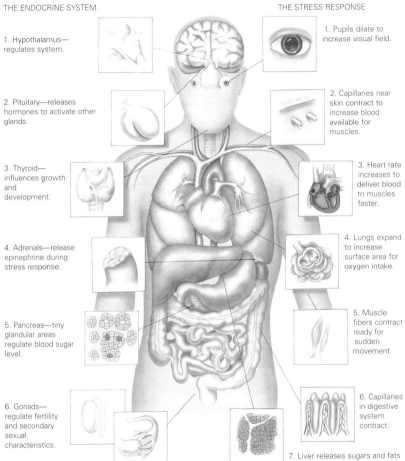

THE ENDOCRINE SYSTEM.

1. Hypothalamus—regulates system.

2. Pituitary—releases hormones to activate other glands.

3. Thyroid—influences growth and development.

4. Adrenals—release epinephrine during stress response.

5. Pancreas—tiny glandular areas regulate blood sugar level.

6. Gonads—regulate fertility and secondary sexual characteristics.

THE STRESS RESPONSE

1. Pupils dilate to increase visual field.

2. Capillaries near skin contract to increase blood available for muscles.

3. Heart rate increases to deliver blood to muscles faster.

4. Lungs expand to increase surface area for oxygen intake.

5. Muscle fibers contract ready for sudden movement.

6. Capillaries in digestive system contract.

7. Liver releases sugars and fats into blood to fuel muscle cells.

ABOVE The chief glands of the endocrine system, and their actions, are shown on the left-hand side of the diagram. One of the fastest-acting reactions is caused by the release of epinephrine (adrenaline) from the adrenal glands in response to stress, reactions whose fundamental purpose is to prepare the body—particularly the muscles for "fight or flight." Within seconds a host of changes take place, as shown on the right-hand side of the diagram.

Endocrine System

Drugs that prevent the production of thyroid hormones are often used in the treatment of Graves' disease. Alternatively, surgery may be carried out to remove part of the gland. Another method consists of treating the patient with a radioactive isotope of iodine, which is taken up by the thyroid and acts to destroy part of it. Both of the treatments that destroy part of the gland tend to result in an underactive thyroid, which may then have to be treated by hormone replacement therapy.

Some people, typically middle-aged women, develop a goiter because of an autoimmune disease in which the body's defense mechanism is muddled. It treats its own tissue as "foreign" and produces antibodies to attack it. In Hashimoto's disease, the thyroid is infiltrated by large numbers of lymphocytes, or white blood cells produced by the body in its confusion.

As with the other endocrine glands, underactivity of the thyroid—or hypothyroidism

—can have effects that are just as serious as those of overproduction. A baby suffering from a deficiency of thyroid hormones may stop growing and, if left untreated, the result can be the severe mental and physical retardation known as cretinism. To prevent this happening treatment has to start as early as possible—thus babies are tested for blood levels of the hormone soon after birth. The test involves measuring the level of thyroid stimulating hormone in a small sample of blood, with excessive levels showing that there may be a problem.

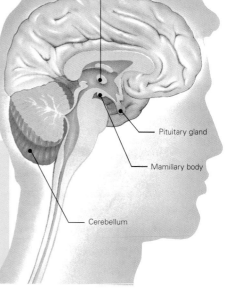

Hypothalamus

Pituitary gland

Mamillary body

Cerebellum

THE HYPOTHALAMUS
The hypothalamus contains neurons that control releases from the anterior pituitary. Seven hypothalamic hormones are released into a portal system connecting the hypothalamus and pituitary, and cause targets in the pituitary to release eight hormones.

Endocrine System

Treatment with thyroxine gives a rapid improvement and means that the baby can develop normally.

In adults hypothyroidism causes myxedema, whose symptoms include tiredness, a hoarse voice, intolerance of cold, dry skin and loss of hair, depression, infertility, and menstrual abnormalities in women. Diagnosis is carried out by analyzing the blood for levels of TSH (which will probably be high) and thyroxine and triiodothyronine (which will be low). The disease can be caused by various conditions, including iodine deficiency, damage to the thyroid, and malfunction of the pituitary. Treatment is by administering thyroxine orally.

BELOW The hypothalamus and the pituitary lie close together and are linked to the brain by nerves and blood vessels. A message from the brain along nerves to the hypothalamus stimulates it to release hormones. Some, such as oxytocin and vasopressin, travel along neurosecretery tracts to be stored in the posterior lobe of the pituitary.

HORMONES RELEASED
The hormones produced and released from the anterior pituitary include: thyrotropin, adrenocotictropin, luteinizing hormone, follicle-stimulating hormone, growth hormone, prolactin, melanocyte-stimulating hormone, endorphins, and enkephalins.

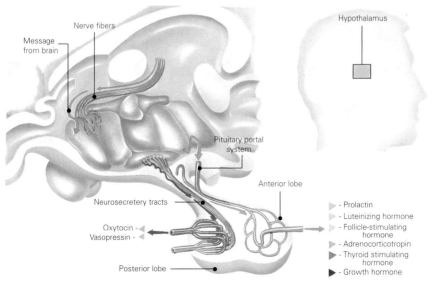

Nerve fibers

Message from brain

Hypothalamus

Pituitary portal system

Anterior lobe

Neurosecretery tracts

- Prolactin
- Luteinizing hormone
- Follicle-stimulating hormone
- Adrenocorticotropin
- Thyroid stimulating hormone
- Growth hormone

Oxytocin -
Vasopressin -

Posterior lobe

Endocrine System

PARATHYROID GLANDS

Calcium is one of the vital minerals. It is mainly used in the formation of bones and teeth, but there is also a small amount of free calcium in the body. It is needed for the brain, nerves, and muscles to function properly and for blood clotting. It may also be involved in the everyday workings of cells. Vital control of the body calcium level is carried out by the hormone parathormone, produced by the parathyroid glands, pea-sized organs located on the back of the thyroid.

When the level of parathormone is increased, stored calcium is released from the bones. Because bones are mainly calcium phosphate, the release of calcium also causes a release of phosphate. To prevent excessive phosphate build-up, the parathormone also increases the rate at which the kidneys excrete phosphate from the body.

Overactivity of the para-thyroids gives high levels of blood calcium while the consequential loss of calcium from the skeleton can soften the bones. Kidney stones can come about because of high blood calcium levels. Conversely, a lack of parathormone—possibly due to damage of the parathyroids—results in an abnormally low level of blood calcium. The main symptom is uncontrolled twitching and spasms of the body muscles, a condition known as tetany, with muscle cramps typifying less severe cases. Increasing calcium intake either in tablet form or by eating calcium-rich dairy products is the best treatment.

To get the calcium from food the body relies on vitamin D, which is processed in the liver and kidneys to make the hormone calcitriol. Output of calcitriol is stimulated by an increase in the level of parathormone and a decrease in the blood phosphate. Calcium levels are critical in growing children, and in women who are pregnant or breast-feeding their babies. To make sure that the calcium needs are met, the production of calcitriol is also stimulated by the growth hormones, estrogen (produced during pregnancy), and prolactin (associated with breast-feeding).

Pituitary gland

Prolactin stimulates milk production

Suckling releases oxytocin

Contraction of milk ducts

LEFT The production of milk after childbirth is triggered by the baby suckling at the mother's breast. This action makes the pituitary release the hormones oxytocin, which squeezes milk from the milk ducts, and prolactin, which maintains the production of milk.

THE MASTER GLAND
The pituitary gland is often called the master gland and is located in a small bony cavity at the base of the brain.

Endocrine System

THE PANCREAS
—A DUAL-PURPOSE GLAND

Active body cells need energy and they get it from carbohydrate food which is broken down into glucose and absorbed into the bloodstream. Glucose gets from the blood and into the cells with the help of the hormone insulin, which is produced in the pancreas. Most of the pancreas is engaged in producing enzymes. When more energy is needed, the glycogen is released from store and converted back to glucose.

To tune the system to perfection the islets of Langerhans also produce glucagon, a second hormone with a complementary action to that of insulin. It comes into play when the glucose level in the blood starts to fall. To prevent the level falling too low, the glucagon stops the production of glycogen and starts to convert the glycogen in the liver back into glucose. Working in harmony, insulin and glucagon thus maintain the blood glucose at the level needed for the body to work normally.

But the system does not always operate with complete smoothness. The disease diabetes mellitus is the result of something going wrong with the making or use of insulin. Glucose builds up in the bloodstream but does not get into the cells. This causes tiredness and leads to weight loss because fat and muscle have to be burned up to provide the necessary energy. Blood glucose levels stay high and the glucose in the bloodstream spills over into the urine, making it characteristically sweet. The rate of urination goes up and is accompanied by extreme thirst.

Sufferers from diabetes may show a wide range of symptoms, and there are

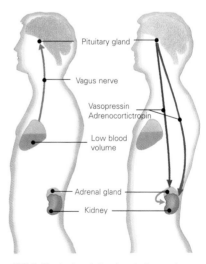

Pituitary gland

Vagus nerve

Vasopressin
Adrenocortictropin

Low blood
volume

Adrenal gland

Kidney

ABOVE Blood volume is largely under hormonal control. When blood volume falls, the vagus nerve "tells" the pituitary, which releases two hormones: vasopressin and adrenocortictropin (ACTH). Vasopressin slows down the release of fluid by the kidneys, and ACTH acts on the adrenals to produce aldosterone, making the kidneys retain salt.

THE PANCREAS
The pancreas contains exocrine cells that secrete digestive enzymes into the small intestine and clusters of endocrine cells (pancreatic islets). These islets secrete the hormones insulin and glucagon.

Endocrine System

many possible complications, which may involve all parts of the body, including the eyes, kidneys, blood vessels, and heart. Most are commonest in diabetics who have had the disease for a long time.

For this reason diabetes is one of the most common causes of blindness, either by causing a cataract (in which the eye lens becomes opaque) or by causing actual damage to the retina. Similarly diabetes can make the arteries narrow and be a trigger of heart attacks. Sometimes the arteries supplying an extremity such as a hand or foot can close off so much that they become blocked. Because the limb gets no blood, gangrene may set in. At worst the affected limb has to be amputated.

In children diabetes is often caused when the pancreas fails to produce enough insulin. It can appear suddenly and cause severe disruption of the body's metabolism. In the past this juvenile diabetes was often fatal, but once insulin had been discovered and isolated, it became possible to treat juvenile diabetes by directly administering insulin to increase the hormone levels in the bloodstream. But insulin is a protein so

you cannot take it as tablets—it would simply be digested and broken down along with other proteins in food. Instead it has to be injected directly into the body. This is a very effective treatment for diabetes, and the insulin used is obtained from the pancreas glands of cows or pigs, or from bacteria that have been genetically engineered so that they produce human insulin.

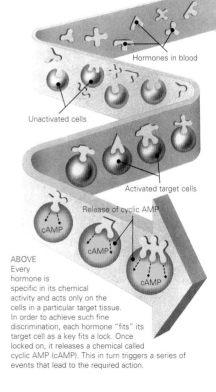

Hormones in blood

Unactivated cells

Activated target cells

Release of cyclic AMP

cAMP

cAMP

cAMP

MENSTRUAL CYCLES
The menstrual cycle is controlled by a number of hormones secreted in a cyclical fashion. Thyroid secretion is usually higher in winter than summer. Childbirth is also hormonally controlled, and is highest between 2 and 7 am.

ABOVE
Every hormone is specific in its chemical activity and acts only on the cells in a particular target tissue. In order to achieve such fine discrimination, each hormone "fits" its target cell as a key fits a lock. Once locked on, it releases a chemical called cyclic AMP (cAMP). This in turn triggers a series of events that lead to the required action.

Endocrine System

ADRENAL GLANDS

Each of the two adrenal glands, found on the upper part of the kidneys, has two main parts, an outer cortex and an inner medulla. These produce a number of different hormones that have a wide variety of functions. The outer part of the cortex produces the mineralocorticoids—steroid hormones that act to control the fluid and mineral balances of the body. Raised blood pressure can be caused by Conn's syndrome, in which over-production of these hormones makes the body hold on to too much sodium and water.

Control of the way the body deals with carbohydrates, particularly glucose, is a vital function of glucocorticoids. They are produced by the two inner layers of the cortex. Male sex hormones—androgens—are also made by the adrenal cortex (even in women, although the main source in men is the testes). Genetic defects may make the adrenal glands produce an excess of androgens in early life, and when this happens to females it tends to generate masculine characteristics. These include enlargement of the clitoris, making it very difficult to establish a child's true sex. In extreme cases such girls can look like boys. Appropriate treatment to block the production of androgen reduces the symptoms and allows the girls to become fertile.

The adrenal glands are also active at the moment of birth. Research has revealed that the birth process may be triggered by androgens made by the adrenal glands of the baby. In the last few weeks of pregnancy the level of output seems to increase and the androgens upset the mother's hormone balance and lead to the onset of labor.

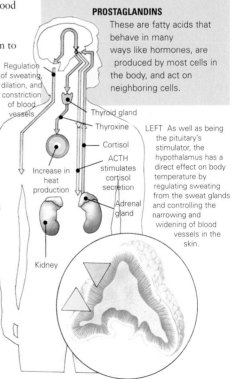

PROSTAGLANDINS
These are fatty acids that behave in many ways like hormones, are produced by most cells in the body, and act on neighboring cells.

Regulation of sweating, dilation, and constriction of blood vessels

Thyroid gland

Thyroxine

Cortisol

ACTH stimulates cortisol secretion

Increase in heat production

Adrenal gland

Kidney

LEFT As well as being the pituitary's stimulator, the hypothalamus has a direct effect on body temperature by regulating sweating from the sweat glands and controlling the narrowing and widening of blood vessels in the skin.

Endocrine System

GONADS—THE SEX GLANDS

Sexual development and activity would be impossible without the action of the gonads—the testes in men and the ovaries in women. The different courses of development typical of a girl or a boy start in the womb when the embryo is about 40 days old. If the embryo is male the embryonic gonads start to develop as testes and begin producing the male sex hormone testosterone. This acts on the, as yet undeveloped, sex organs and they start to form a penis and scrotum. With a female embryo there is no male hormone and the sex organs form the Fallopian tubes, uterus, cervix, and vagina, while the gonads become ovaries.

When a baby is born its ovaries or testes are not yet working at full capacity, but they do produce small amounts of hormones—mainly estrogen in girls and testosterone in boys. Similarly the pituitary gland produces minute amounts of its sex hormones but the system is really doing little more than ticking over. Then as puberty approaches the picture changes dramatically, with a sudden surge of sex hormones. Exactly what causes the changes is not certain but the results are obvious. Girls start to develop breasts and

their hips widen as they turn into women; boys start to grow facial and body hair, and their voices deepen as they become men; and in both sexes underarm and pubic hair starts to grow.

These secondary sexual characteristics start to show because of the action of the sex hormones—estrogen, progesterone, testosterone, and androsterone. The first two are primarily associated with female sexual functions and the second two with male sexual functions, but all four are produced by both males and females. The different characteristics are the result of variations in the relative levels of the hormones.

Puberty normally occurs between the ages of 10 and 16 in girls, and 14 and 18 in boys, but some conditions can cause sexual development to occur before or after these ages. For example, if the

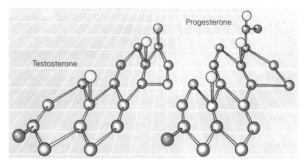

BELOW Most of the difference between males and females—established initially at the embryonic stage in the womb—result from the presence of sex hormones, chief of which are testosterone in males, and estrogen and progesterone in females. The difference in molecular structure of testosterone and progesterone is very little.

Progesterone

Testosterone

Endocrine System

hypothalamus or pituitary are damaged or affected by a tumor there may be a premature release of the gonadotropins. Abnormally early development of the gonads takes place, with the children developing full sexual maturity and the appropriate secondary sexual characteristics. In contrast abnormalities in the testes or ovaries may restrict the production of the sex hormones, and delay the onset of puberty.

From their first menstrual period (menarche) onward, women have a regular, hormone-controlled, reproductive cycle that prepares their bodies for fertilization and childbearing. This menstrual cycle works through every 28 days or so, with ovulation—the release of an ovum—taking place about halfway through it. The womb lining is shed as the menstrual flow at the end of each cycle. All of the hormonal changes involved are controlled by the gonadotropins secreted by the pituitary.

During the first part of the menstrual cycle, follicle-stimulating hormone has a dominant effect. Its action causes one of the ovarian follicles to become dominant and fully develop into a Graafian follicle. Estrogen is released by the follicle and acts to prepare the lining of the womb (the endometrium) to accommodate a fertilized ovum. About 14 days into the cycle there is a surge of luteinizing hormone to provoke the release of the ovum by the ripe follicle, which becomes the corpus luteum. This organ then secretes progesterone, which acts with the estrogen

to build up the lining of the womb.

Progress of the rest of the cycle now depends on what happens to the released ovum. If fertilization does not occur within 48 hours of release, the ovum dies. The corpus luteum degenerates, and secretion of progesterone decreases, from the seventh to the fourteenth day after ovulation. Reduction in the hormone levels makes the womb lining break down and it is shed by menstrual bleeding. The start of menstruation is taken as the beginning of a new cycle.

A different sequence of hormone production is set into operation if an ovum happens to be fertilized by a sperm from a man. About four days after fertilization the developing embryo burrows into and embeds itself into the lining of the womb and starts to form a placenta. The newly forming placenta releases a hormone that acts on the corpus luteum of the ovary so that it continues to produce progesterone and the womb lining is kept and maintained. When developed, the placenta takes over the functions of the corpus luteum to produce most of the steroid hormones needed, as well as being vital to the nourishment of the developing child.

INTERFERONS

These are proteins released when a cell is attacked by a virus. They cause neighboring cells in the body to produce antiviral proteins, and once activated, these proteins destroy the virus.

The
BODY REGIONS

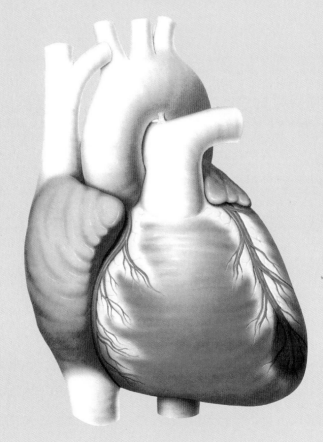

As well as various body systems enabling us to function properly, there are certain very important organs. Without them we would not be able to live a healthy and fulfilling lifestyle. In this section of the book we look at body regions.

CELLS

The cell is the smallest unit of living matter. Every type of tissue in the body, from bones and brain to muscles and skin, is made up of cells. And the sperm and the egg—those microscopic scraps that come together to form a new human being—are each only a single cell.

SKIN

The skin is the largest organ of the body and is made up of a thin outer layer called the epidermis, and a thicker outer layer called the dermis. Below the dermis is the subcutaneous tissue, which contains fat. Buried within the skin are sebaceous glands that secrete a lubricating substance called sebum. There are also nerves that sense cold, heat, pain, pressure, and touch along with sweat glands, which produce perspiration when you are too hot.

BRAIN

The brain is one of the largest and most important organs of the human body. This organ has a wide range of responsibilities from coordinating our movements to managing our emotions. The brain is made up of three main parts; the forebrain, the brainstem, and the hindbrain. These three parts collectively weigh in at about three pounds.

HEART

The heart is one of the major parts of our body. It is as small as your fist but does a major job in our body. The heart delivers blood throughout our bodies, by arteries, capillaries, and veins. As the heart is a strong, muscular pump that continuously works around the clock, it is important for us to understand how it works and to take care of it. In an average lifetime, the human heart beats more than 2.5 billion times.

KIDNEYS

The contemplation of physical well-being tends to involve thinking mainly of inputs: getting enough oxygen, food, and fluids. The less appealing matters of waste disposal do not receive such attention, but they are just as vital, for without them the body is liable to be poisoned by itself.

SENSES

You can only know of the existence of an object—we can only prove it is there—because we can see it, hear it, taste it, smell it, or feel it. You would be totally unaware of the presence of other people—or indeed of anything else in the world around you—if it were not for your senses.

CELLS

Cells group together to form tissues, and tissues in turn form organs and body systems. Each type of tissue has its own type of cell, and as a result there are hundreds of different kinds of cells in the body. Superficially a typical cell resembles a blob of jelly in a membranous bag. But it can replicate and sustain itself within its environment and it can grow. It takes in what it needs to function, and produces other substances that the body needs to stay alive.

A cell therefore consists of the essential components needed for the life process, contained within its enveloping membrane. These essentials are the chemicals and enzymes which react to provide the cell with energy and building materials, together with genetic material which bears the information for producing cell components and for cell replication. The cell is like a factory which is supplied with raw materials and processes them to make products. The products themselves are used either to make components for the factory's production lines or are transported out of the cell to be used elsewhere.

TWO KINDS OF CELLS

Essentially, there are two types of cells—procaryotic cells and eucaryotic cells. The two differ fundamentally in the way that they are organized within. In procaryotic cells, the arrangement is simple. The chemicals and enzymes needed to produce energy, and those needed for cell growth and division, are contained in the cytoplasm—a complex jellylike mixture—which is packaged in a plasma membrane that forms the cell boundary. These cells do not have a nucleus, and their genetic

materials—deoxyribonucleic acid (DNA)—is attached to the plasma membrane. Procaryotic cells were probably the first sort of cells to appear on Earth. Most procaryotes today are single-celled organisms, notably bacteria and blue-green algae, and none is found in the human body—or in any other living animal.

Eucaryotes, on the other hand, are much larger than procaryotes, and have a far more highly evolved and complicated internal arrangement. They are typical of the cells in the human body. In them, the genetic material is contained within a nucleus, which is surrounded by nucleoplasm and bounded within its own membrane. Surrounding this nuclear membrane is the cytoplasm.

Generally, eucaryotes are highly specialized cells able to perform specific functions by the action of internal

PROCARYOTES
These are cells without a nucleus, which include bacteria and cyanophytes. The genetic material is a singular circular DNA and is contained in the cytoplasm, since there is no nucleus.

Cells

structures called organelles. These intracellular compartments, usually within the cytoplasm, coordinate cell chemistry. They also make essential cell chemicals and export the products manufactured by the cell. The organelles, by concentrating cell components together, can make the cell's internal biochemical reactions more efficient, and can "package" potentially dangerous chemicals, effectively isolating them and preventing them from poisoning or destroying the cell from within.

EUCARYOTES
These cells do have a nucleus and are found in humans and other multicellular organisms (plants and animals), and also algae, and protozoa. They have both a cellular membrane and a nuclear membrane.

BELOW The skin of the fingers has a unique pattern of whorled ridges and grooves. Cells in the subcutaneous tissue just below the skin's surface die and are sloughed off as they reach the surface.

Cells

A PLAN FOR LIFE

The central nucleus is the largest and most prominent of the many organelles in a human body cell. It contains all the information needed by the cell to carry out its functions—the manufacture of components and self-replication. It houses the blueprint or program for the life process.

Within the nucleus is the nucleolus—a structure involved in protein manufacture via ribonucleic acid (RNA), which works in collaboration with DNA—and the nucleoplasm, which is the cell's "reference library." This file of genetic information appears in the form of DNA packaged into a number of chromosomes. These chromosomes are built of proteins, called histones, combined with tightly packaged coils of DNA; the combination is called chromatin. The long DNA molecule is spooled twice around sets of eight histones to form a nucleosome, and numerous nucleosomes wound with the same DNA strand form a chromosome. When cells multiply, or replicate themselves, genetic information is passed on in exactly the same form. Just before mitosis (cell division), the chromosomes

SURROUNDED BY A PLASMA MEMBRANE
All are surrounded with a membrane, and within this is the cytoplasm which is composed of the fluid and organelles of the cell.

are duplicated. This means that the two daughter cells created when the cell divides receive an identical copy of each of the cell's chromosomes.

The nuclear membrane—the double-walled structure that envelops the nucleus—is perforated with submicroscopic holes. This arrangement allows a two-way flow of molecules between the nucleoplasm in the nucleus and the cytoplasm immediately outside it. In this way, proteins needed to build the nucleus and keep it working enter the nucleus from the cytoplasm, where they are processed. In return, the RNA essential for protein synthesis passes from the nucleoplasm out into the cytoplasm.

BRICKS OF LIFE

Other types of organelles—ribosomes—are essential for protein synthesis. Ribosomes, which look like minute spheres within the cell, line up on the endoplasmic reticulum, a membrane system enfolded through the cytoplasm. The ribosomes are composed of RNA and protein, which in turn is made up of amino acids linked in chains.

Proteins are vital to life. They perform most of the cell's biochemical reactions and are the structural material from which the cell itself is made. So they function both as the toolkit and framework of the cell. What distinguishes one protein from another is the sequence in which its amino acids are linked. This, in turn, is determined by the coding of the

Cells

DNA chromosome, which directs their manufacture.

In protein synthesis, first the DNA of a gene is copied into RNA. This passes out of the nucleus into the cytoplasm. There ribosomes, enzymes, and other RNAs come together to translate the gene's RNA structure into the amino acid sequence of a specific protein.

BACTERIA
These are procaryotes and have DNA, but it is not organized into a true nucleus with a nuclear envelope around it. Also many other internal organelles like mitochondria and chloroplasts are abscent.

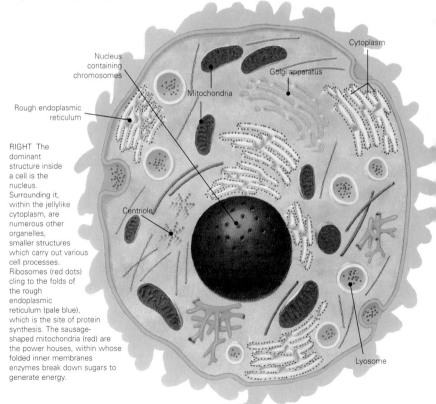

Nucleus containing chromosomes

Mitochondria

Cytoplasm

Golgi apparatus

Rough endoplasmic reticulum

Centriole

Lyosome

RIGHT The dominant structure inside a cell is the nucleus. Surrounding it, within the jellylike cytoplasm, are numerous other organelles, smaller structures which carry out various cell processes. Ribosomes (red dots) cling to the folds of the rough endoplasmic reticulum (pale blue), which is the site of protein synthesis. The sausage-shaped mitochondria (red) are the power houses, within whose folded inner membranes enzymes break down sugars to generate energy.

Cells

In this complex process, three kinds of RNA molecules interact during synthesis. Messenger RNA (mRNA) "read" from the DNA provides the coding for the amino acid sequence, transfer RNA (tRNA) carries the amino acids that are being assembled, and ribosomal RNA is bound into particles that link the amino acids to one another. The ribosomes have three slots on their surface: one holds the mRNA and the other two hold the tRNAs while enzymes bound to the ribosomes link the amino acids in the coded sequence to form the protein.

THE ENGINE OF THE CELL
Mitochondria are found in nearly all eucaryotic cells, usually several or many per cell, and these burn sugar for fuel in the process of cellular respiration—they're the "engine" of the cell.

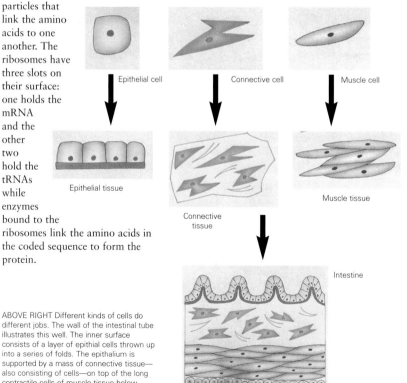

Epithelial cell

Connective cell

Muscle cell

Epithelial tissue

Connective tissue

Muscle tissue

Intestine

ABOVE RIGHT Different kinds of cells do different jobs. The wall of the intestinal tube illustrates this well. The inner surface consists of a layer of epithial cells thrown up into a series of folds. The epithalium is supported by a mass of connective tissue—also consisting of cells—on top of the long contractile cells of muscle tissue below.

Cells

SYNTHESIZER AND SORTER

The endoplasmic reticulum, on which the ribosomes are clustered, has two important jobs to do in the cell. It helps in the synthesis of large molecules such as proteins, lipids, and complex carbohydrates that make up some of the other organelles in the cell. And it also separates newly synthesized molecules needed by the cytoplasm from those intended for transport to other sites. It consists of a membrane folded in the form of a crumpled egg-shaped sac that surrounds the nuclear envelope. This enclosure is the cisternal space, or lumen.

The folds of the membrane farthest from the nuclear envelope form a boundary in the cytoplasm. Proteins are assembled on the cytoplasmic side of this boundary and must pass through it into the cisternal space if they are to be exported. The endoplasmic reticulum also separates potentially dangerous proteins, such as digestive enzymes, from the rest of the cell as they are made.

Studies of the cell using an electron microscope reveal two distinct forms of endoplasmic reticulum. One is rough or granular, with ribosomes attached to its outer surface, and is a site of lipid synthesis. Smooth endoplasmic reticulum is composed mostly of fine hollow tubes, unlike the rough form, which has a flattened sac-shaped interior.

The amounts of rough and smooth endoplasmic reticulum in any cell depend on what the cell has to do in the body. Cells that make large quantities of proteins for export, such as B-lymphocytes, in which antibodies are formed, have a large proportion of the rough form, whereas cells that specialize in processing lipids—for instance in the production of steroid hormones—have much more of the smooth type.

Proteins and membranes made by the endoplasmic reticulum are transported to their destination by an ingenious mechanism in the cell—in "buds" or vesicles that sprout from the endoplasmic membrane. If the cell surface is the intended destination, then the buds fuse with it and release the newly formed substance. If a cell organelle is the destination, then the bud fuses with it and the substance is released inside it.

MITOCHONDRIAL DNA
While half of the DNA in the nucleus of the newly formed embryo comes from the mother and half comes from the father, and because sperm does not pass any of their mitochondria to the offspring, the mitochondrial DNA comes only from the mother.

Cells

PACKAGING PROTEINS

Many of the vesicles containing substances synthesized by the endoplasmic reticulum are destined for the Golgi apparatus—a stack of interconnecting membranes and spaces near the nucleus. This miniature packaging factory within the cell is probably important in the final processing of the newly synthesized proteins that reach it. After processing, the protein is transported, also in vesicles, to its destination.

The Golgi apparatus and the endoplasmic reticulum manufacture two organelles—lysosomes, responsible for digestion inside the cell, and peroxisomes, the cell's detoxification plants, which break down alcohol and other potential poisons to prevent them from killing the cell. The packaging of substances into "cartons," or vesicles, is important to cell safety. Both digestion and detoxification involve powerful chemicals, such as enzymes and oxygen, which would damage the rest of the cell should they be free to come into contact with it.

WITHOUT EUCARYOTES
Although eucaryotes use the same genetic code and metabolic processes as procaryotes, without eucaryotes, the world would lack mammals, birds, fish, invertebrates, mushrooms, plants, and complex single-celled organisms.

BELOW and OPPOSITE Particles passing in and out of cells are packaged for this purpose. Infolding of the double-layered cell membrane during endocytosis allows food particles to enter the cell's cytoplasm from the fluid exterior. The membranous fold gradually encircles the particles and forms a vesicle in which they are transported within the cell. Particles of waste products are first packaged in a vesicle and transported to the cell wall, where the reverse process of exocytosis enables them to pass through the membrane and out of the cell.

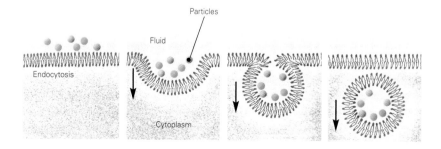

Particles

Fluid

Endocytosis

Cytoplasm

Cells

THE DIVIDING LINE

All human body cells are bounded by an outer plasma membrane. It contains the cell and isolates the contents from the environment. But equally important, the membrane is a wall with doors in it. By opening and closing its doors, it regulates the passage of substances into and out of the cell. The plasma membrane is only about one four-thousandth of an inch thick. And its structure is essentially the same in all living organisms. It is composed of proteins sandwiched between two layers of fatty phospholipids.

The phospholipids can be pictured as tadpole-shaped molecules with hydrophobic (water-hating) heads and hydrophilic (water-loving) tails. Thousands of the "tadpoles" line up side by side and tail to tail on two layers to form the plasma membrane. Scattered among the ranks of phospholipids are "islands" of protein material which act as "doors" in the membrane wall. All interactions between the environment and

the cytoplasm inside the cell must involve this membrane. Molecules outside the cell may react with membrane proteins. This opens the "doors" in the wall and makes them suitable to cross the membrane. The proteins may also bind the cell to its neighbors, so that they can work together in unison.

Some small molecules, such as oxygen and carbon dioxide, can cross the plasma membrane without opening doors. They do so simply by diffusing from a region of high concentration to one of low concentration. Others, such as glucose or sodium and potassium ions, need an active transport process—the doors open only if there is energy available.

THE SIZE OF EUCARYOTES
Eucaryotic cells are about ten times the size of a procaryote and can be as much as 1,000 times greater in volume.
It is the nucleus of this cell that gives the eucaryote—literally, true nucleus—its name.

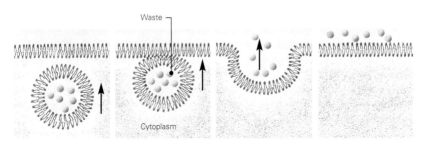

Waste

Cytoplasm

Cells

BELOW Dozens of different kinds of cells form the various body tissues. All except red blood cells share the common feature of a central nucleus (shown green), but their shapes can be very dissimilar. The neurons, or nerve cells, in the brain have many tendril-like branches to make contact with neighboring nerve cells, whereas bone cells amass mineral particles locked into a rock-hard matrix.

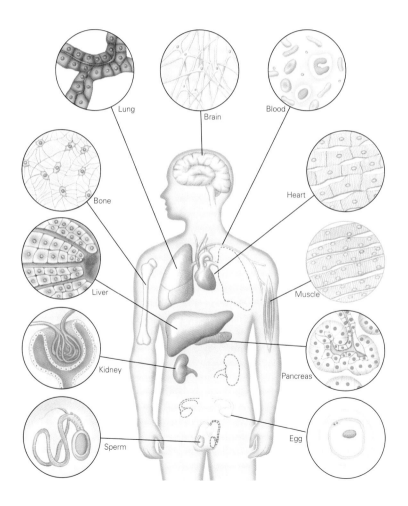

Lung

Brain

Blood

Bone

Heart

Liver

Muscle

Kidney

Pancreas

Sperm

Egg

Cells

The sodium-potassium pump by which molecules of ATP (adenosine triphosphate) are broken down to release energy, which in turn controls the amount of salt inside a cell, is just such a process.

Molecules may also open the doors in the plasma membrane by means of a special cell vehicle. One such vehicle is created as a dimple, which forms in the membrane and engulfs the incoming substance—perhaps a bacterium, which is "eaten" by the cell and destroyed. This is phagocytosis. Another similar mechanism is pinocytosis, the "eating" of fluid by the membrane.

THE GREAT DIVIDE

Growth, renewal, and repair are fundamental features of all life. All are made possible by mitosis—the process by which cells reproduce themselves and pass on their characteristics to the next generation. The genetic information is carried in DNA (deoxyribonucleic acid), the molecule of life which, together with protein, makes up the bulk of the nucleus. The genes are strung on to the chromosomes like beads on a necklace. But just how the genes might be handed down remained a mystery until the mid-twentieth century, when researchers James Watson, Francis Crick, Maurice Wilkins, and Rosalind Franklin, working at Cambridge University in England, identified the DNA molecule and explained its structure and function.

DNA is a giant molecule in the form of a double helix resembling a twisted rope ladder. Each side of the ladder is a single strand made up of alternative sugar and phosphate molecules. The sugars are the points where the "rungs" meet the strand and the phosphates are the rope between the rungs. The rungs of the ladder are formed by pairs of nitrogen-containing bases linked to the sugars and joined by hydrogen bonds.

Each unit of the helix, comprising a sugar, a phosphate, and a base—equivalent to a knot, the rope between knots and the half-rung—is a nucleotide. There are four nitrogen bases—adenine and thymine, and cytosine and guanine—which always pair to form the rungs. The number of nitrogen bases on the helixes number many thousands. It is their exact order and sequence that form the genetic code, the vital information which controls the synthesis of various different proteins in the cell.

Before the cell divides, the cell contents go through a preparatory stage, during which the DNA replicates or forms identical copies of itself.

THE NUCLEUS
The nucleus is the most conspicuous organelle found in the eucaryotic cell. It houses the cell's chromosomes and is the place where almost all DNA replication and RNA synthesis occurs.

Cells

In this remarkable natural replication mechanism, the twisted rope ladder arrangement separates lengthways as the hydrogen bonds linking the half-rungs break. On each strand, nucleotides within the nucleus attach to the bases to rebuild the structure, and so two molecules of DNA are formed. They are identical because the order of the bases is retained on each newly separated strand, and the bases always pair with the same partner.

After DNA replication, the nuclear membrane breaks down and the DNA can be seen within the cell as distinct strands wrapped in protein. These are the chromosomes, each of which contains two strands of DNA. The strands divide into two and each moves to opposite sides of the cell. As the cell divides, each daughter cell receives an identical component of DNA with identical coding. Then the strands "unravel" in the nucleus of the new cells. They continue their essential work within the cell, but are much more difficult to discern as individual entities.

LYSOSOMES AND PEROXISOMES
These are often referred to as the garbage disposal system of a cell. Both are spherical, bound by a single membrane and rich in digestive enzymes.

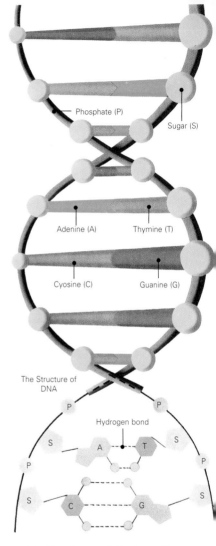

Phosphate (P)

Sugar (S)

Adenine (A)   Thymine (T)

Cyosine (C)   Guanine (G)

The Structure of DNA

Hydrogen bond

RIGHT Three key components combine to form the long double helix molecule of DNA, full name deoxyribonucleic acid. The sides of the strands, resembling a twisted ladder, are formed of alternate sugar (pale green) and phosphate (purple) groups joined in long chains. Each "rung" of the ladder consists of two nitrogen-containing bases, either adenine (dark green) and thymine (orange) or cytosine (red) and guanine (blue). The bases are held together at mid-rung by comparatively weak hydrogen bonds. DNA makes up the genetic material of every cell, located on chromosomes inside the nucleus.

Cells

FROM STRENGTH TO STRENGTH

Just as cells may take in food particles by first enveloping them in the cell wall or membrane, so, by a reverse process, waste products packaged in similar compartments are shipped to the membrane. The door-opening mechanism out of the cell works like this: the walls of the vesicle fuse with the membrane to form an opening to the outside of the cell, and the waste is ejected like garbage out of a rubbish chute.

Once inside the cell, food is put to good use. It is broken down in the cell's mitochondria to provide energy for protein synthesis, waste disposal, cell replication, the movement of cell components, and other cell activities. The mitochondria contain reaction-speeding enzymes—proteins made in the cell which are crucial to the metabolic process.

RIGHT DNA unzips down the middle when it replicates during cell division. The "rungs" of the parent helical ladder (lower part of diagram) break at their weak hydrogen bonds, forming two "half ladders" (center of diagram). Free nucleotides—a base attached to a sugar and a phosphate—come along and attach to their usual partner, reforming the rungs and creating two new and identical daughter DNA strands (top).

DNA PROTECTION
The membrane of the cell called the nuclear envelope, isolates and protects the cell's DNA from various molecules that could accidently damage its structure or interfere with its processing.

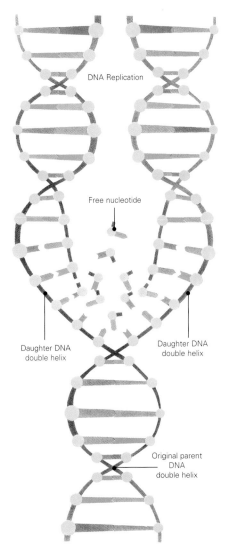

DNA Replication

Free nucleotide

Daughter DNA double helix

Daughter DNA double helix

Original parent DNA double helix

Cells

Without these the biochemical reactions that break down the cell's food would be too slow to be effective. And enzymes are specific—they act on only one type of chemical change. The breakdown of a simple sugar involves more than 50 reactions, each with its own specific enzyme.

These energy-generating reactions within the cell are known collectively as respiration (sometimes called internal respiration to distinguish it from the external respiration that takes place in the lungs). The respiratory reactions occur in two stages. In the first stage, food is only partly broken down into intermediate substances, such as alcohol and acids. This stage does not require oxygen and is called anaerobic respiration. The second stage, aerobic respiration, can only happen if oxygen is available. In it, the intermediate substances are broken down completely, into the waste products carbon dioxide and water, and energy essential to all life is released.

RIGHT Production of proteins within a cell also involves the unzipping of DNA. The sequence of bases on the unzipped strands acts as a code and is reproduced as messenger RNA (mRNA), which attaches itself to a ribosome. Molecules of transfer RNA (tRNA) with the complementary sequence of bases arrive, each locked on to one of 20 or so different amino acids. The acids separate from the RNA and link to each other to form a polypeptide chain, the structural unit of a protein.

THE RIBOSOME
Found in both procaryotes and eucaryotes, there may be hundreds or even thousands of them in a single cell. They are responsible for processing the genetic instructions carried by a mRNA.

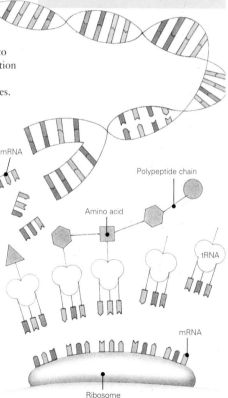

DNA

mRNA

Polypeptide chain

Amino acid

tRNA

mRNA

Ribosome

Cells

STORAGE UNITS

The energy generated by respiration is not immediately available for powering the cell and its processes. It is placed in temporary storage—as the energy in chemical bonds—and released only as it is needed. The storage medium is the chemical adenosine triphosphate (ATP). This crucial chemical is made during respiration from another essential chemical called adenosine diphosphate (ADP).

As their chemical names suggest, ADP has two phosphate groups (linked to the substance adenosine), whereas ATP has three. The energy from respiration goes to force a phosphate group to bond with an ADP molecule, forming an ATP molecule, which can be stored. This bonding is energy-intensive, but the bonds are easily broken to release the stored energy that went into their formation.

As ATP changes to ADPT and the high-energy bond is broken, energy is released virtually instantaneously, so the cell can react rapidly to energy demands without having to go through the 50 or so reactions involved in respiration. Because the breaking of each phosphate bond releases a specific amount of energy, ATP also provides a precise metering of the amount of energy released to the cell. So there is no wasted energy which might cause heat and burn the cell, and the metabolic processes are not slowed down by being starved of energy. Another advantage of the system is that the phosphate energy bonds are transferable. They can be moved from the ATP account to other substances to increase their energy reserves without any leakage of "funds" from the cell.

SINGLE CELLS

Single cells can make up complete living organisms, such as bacteria. The spherical type are known as cocci, and exist as pairs (diplococcus), bunches (staphylococcus) or chains (streptococcus).

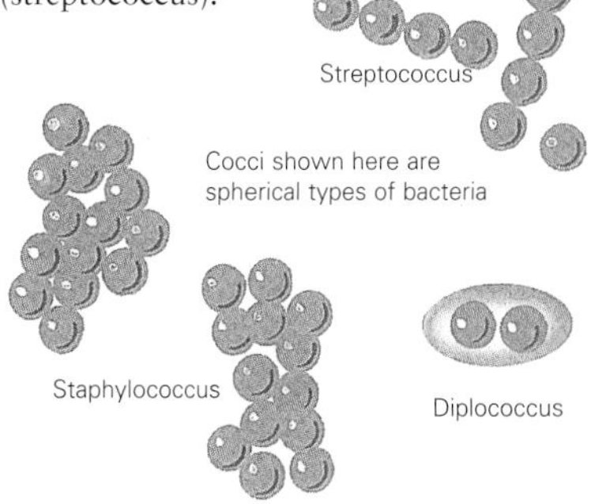

Streptococcus

Cocci shown here are spherical types of bacteria

Staphylococcus

Diplococcus

MITOCHONDRIA
These are self-replicating organelles that occur in various numbers, shapes, and sizes in the cytoplasm of all eucaryotic cells. They also have two functionally distinct membrane systems separated by a space: an outer membrane, which surrounds the whole organelle; and the inner membrane, which is thrown into folds or shelves that project inward.

Cells

Rod-shaped bacteria, with or without flagella as "tails," include those that cause tuberculosis and tetanus. Wormlike forms, some with cilia "whiskers," include salmonella and the organism that causes syphilis.

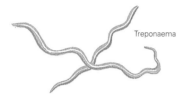

Treponaema

Wormlike bacterial forms, some with cilia "whiskers," are shown here in green.

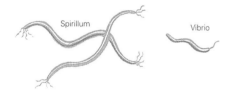

Spirillum

Vibrio

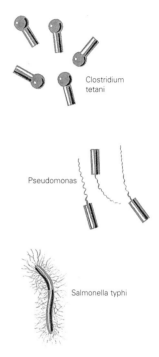

Clostridium tetani

Pseudomonas

Salmonella typhi

Rod-shaped bacteria, with or without flagella as "tails," are shown here in blue.

BINARY FISSION
Bacteria reproduce through a fairly simple process called binary fission, or the reproduction of a living cell by division into two equal, or near equal parts. This type of asexual reproduction theoretically results in two identical cells, however, bacterial DNA has a relatively high mutation rate.

Mycobacterium tuberculosis

Cells

Before a virus has entered a host cell, it is called a viron—a package of viral genetic material. Virions—infectious viral particles—can be passed from host to host either through direct contact or through a vector, or carrier. Inside the organism, the virus can enter a cell in various ways. Bacteriophanges (bacterial viruses) attach to the cell wall surface in specific places. Once attached, enzymes make a small hole in the cell wall, and the virus injects its DNA into the cell. Other viruses (such as HIV) enter the host via endocytosis, the process whereby cells take in material from the external environment. After entering the cell, the virus' genetic material begins the destructive process of taking over the cell and forces it to produce new viruses.

OBLIGATE INTRACELLULAR PARASITES
Because viruses are acellular and do not use ATP, they must utilize the machinery and metabolism of a host cell to reproduce. For this reason, viruses are called obligate intracellular parasites.

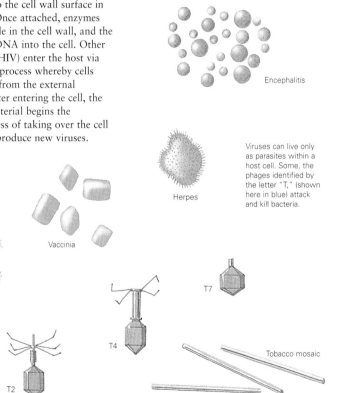

Encephalitis

Herpes

Viruses can live only as parasites within a host cell. Some, the phages identified by the letter "T," (shown here in blue) attack and kill bacteria.

Influenza

Vaccinia

T7

T4

T2

Tobacco mosaic

Cells

CELL DUPLICATION

Cell duplication, the basis of growth and repair, is called mitosis. Its key stages are illustrated on the opposite page. At the beginning of the prophase the duplicated chromosomes become visible in the nucleus and locate themselves on spindle fibers. At metaphase they align across the center of the cell, and in early anaphase they each split in two. Late anaphase sees the chromosomes move apart toward the centioles, as the cell prepares for telophase by constricting across the middle. Finally it splits completely, so that each daughter cell is an exact replica of its parent cell. During interphase the DNA in the two new nuclei replicate, ready for another division.

MAMMILIAN CELL TYPES
Three basic categories of cells make up the mammillian body: germ cells, somatic cells, and stem cells. Each of the approximately 100,000,000,000,000 cells in an adult human has its own copy, or copies of the genome, with the only exception being certain cell types that lack nuclei in their fully differentiated state, such as red blood cells.

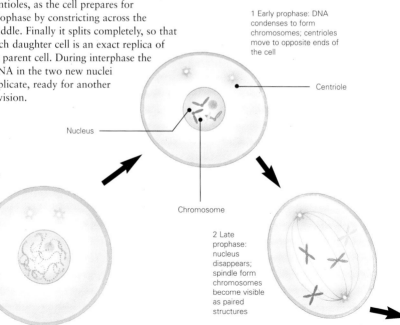

1 Early prophase: DNA condenses to form chromosomes; centrioles move to opposite ends of the cell

Centriole

Nucleus

Chromosome

2 Late prophase: nucleus disappears; spindle form chromosomes become visible as paired structures

Cells

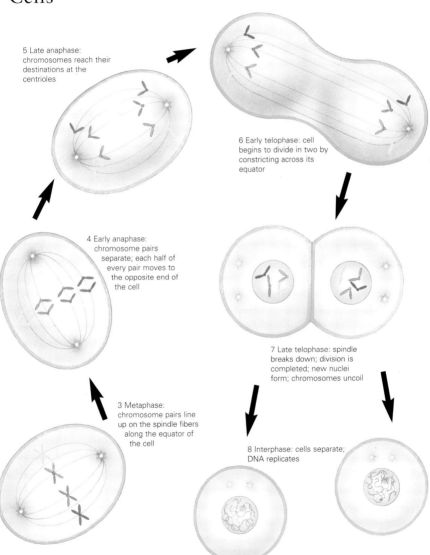

5 Late anaphase: chromosomes reach their destinations at the centrioles

6 Early telophase: cell begins to divide in two by constricting across its equator

4 Early anaphase: chromosome pairs separate; each half of every pair moves to the opposite end of the cell

3 Metaphase: chromosome pairs line up on the spindle fibers along the equator of the cell

7 Late telophase: spindle breaks down; division is completed; new nuclei form; chromosomes uncoil

8 Interphase: cells separate; DNA replicates

Cells

CANCER

All cells need a supply of blood, carrying nutrients and oxygen. But fast-growing tumor cells need more blood than do normal cells.

APOPTOSIS

As cells cycle through interphase and mitosis, a surveillance system monitors the cell for DNA damage and failure to perform critical processes. If this system senses a problem, a network of signaling molecules instructs the cell to stop dividing. These so called "checkpoints" let the cell know whether to repair the damage or initiate programed cell death, a process called apoptosis.

LEFT and BELOW In order for fast-growing tumor cells (shown in red) to obtain more blood supply, they release a substance called tumor angiogenesis factor (TAF, below left), which stimulates the growth of secondary blood vessels at the site of the tumor (below right). The increased nutrient supply allows the cancer cells to grow even more rapidly, setting up a vicious circle of demand and supply.

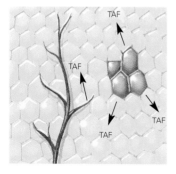

Cells

Cancer occurs when tumor cells multiply out of control and, if unchecked, spread through the body by the process of metastasis. For some as yet unknown reason, cells replicating faster than natural wastage requires gives rise to a primary tumor, which after it exceeds a certain size begins to shed cancer cells. These travel in the bloodstream or through the lymphatic system, where they initiate the growth of secondary tumors at other sites in the body. A primary tumor in the lung, for example, may be tolerable for a while and operable if discovered. But secondary tumors in critical organs such as the liver and brain are frequently the cause of death from cancer.

BELOW Cancer spreads through the body through a process called metastasis.
A Cells replicate faster than normal
B Primary tumor forms
C When the tumor reaches a certain size it starts to shed cancer cells
D Secondary sites develop at other sites in the body

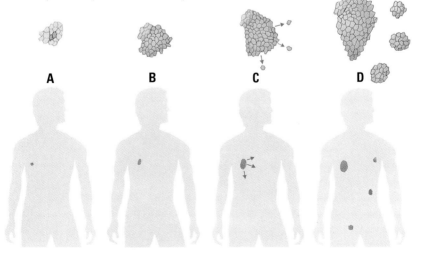

A   **B**   **C**   **D**

SKIN

Silent and immobile, skin registers sensation constantly and supports a teeming, unseen population of tiny organisms. It is adapted to its various functions with remarkable versatility. Not only does it harden from use, but it molds into different shapes. And as it responds to the most delicate touch, skin becomes an organ of communication, sometimes more eloquent than words.

Knitted together with tough cells, skin is the first line of the body's defenses against disease-causing invaders. Repelling countless microorganisms it protects the soft tissue within the body. Nevertheless, its durable exterior is not an impassable barrier. By allowing water and heat to permeate its fabric, the skin helps control the body's temperature.

Skin is a living boundary that separates the inside of the body from the outside world. Constantly in contact with its surroundings, skin is tough enough to resist countless chemical and environmental assaults, yet soft and sensitive enough to respond to the gentlest touch. A versatile organ, skin regulates the movement of substances from the interior to the exterior. Our bodies are made mostly of water—as much as 75 percent at birth and somewhat less later in life. Skin protects this bodily content from a considerably drier environment.

OUTER SKIN CELLS
Humans shed and regrow outer skin cells about every 27 days—almost 1,000 new skins in a lifetime.

The skin barrier is selectively penetrable. Skin secretes fluids that lubricate, barricade toxic substances, and maintain a stable internal environment. It absorbs other substances, particularly those soluble in oils. This absorptive function has proved effective in administering certain medications.

A BARRIER TO THE WORLD

The skin totals between 12 and 20 square feet in area and accounts for 12 percent of body weight. It is composed of three integrated layers—the epidermis, dermis, and subcutis. The terms refer to the positions of these layers as "overskin," "skin," and "underskin." The epidermis is outermost, forming an overall protective covering for the whole body.

The epidermis is never more than one twenty-fifth of an inch thick. Like the other layers, its thickness varies over different parts of the body. It is thickest on the palms of the hands and soles of the feet, where friction is needed for gripping and walking. It is thinnest on the eyelids, which might be light and flexible. The most protean of the layers, the epidermis, also grows into fingernails, toenails, and hair.

As in all tissue, the skin's basic unit is

Skin

the cell. The epidermis is woven of three kinds of cell—keratinocytes, melanocytes, and Langerhans cells. The first synthesize the protein keratin, an essential ingredient of outer skin, hair, and nails. Melanocytes inject granules of pigment into the neighboring cells. Melanin, the dark pigment produced, protects the skin by absorbing the sun's ultraviolet radiation. This radiation can damage the genetic make-up of skin cells, causing skin cancer.

SHEDDING SKIN

We humans shed and regrow about 600,000 particles of skin every hour—about 1.5 pounds a year. By 70 years of age, an average person will have lost 105 pounds of skin.

SKIN FUNCTIONS INCLUDE:

PROTECTIVE—Skin guards the body from harmful environmental elements, including solar radiation, weather, etc.

IMMUNOLOGICAL—Skin prevents entry of microorganisms and actually disposes of harmful bacteria.

FLUID AND PROTEIN BALANCE—Skin prevents fluid loss and distributes essential nutrients to the body.

THERMOREGULATION—Skin prevents heat loss. But it also allows for rapid cooling during exercise through perspiration (sweating) and the vasodilation (widening) of blood vessels.

NEUROSENSORY—Skin possesses nerve endings and receptors which enable the nervous system to process and interpret information (pain, touch, hot, and cold) from the environment.

METABOLISM—Skin is necessary for the production of valuable vitamin D.

SOCIAL-INTERACTIVE—Skin aids in certain social, interpersonal reactions.

Melanin darkens skin exposed to sunlight. The pigment also accounts for different colors of skin, hair, and eyes, Langerhans cells aid the body's immune system by intercepting alien material in the skin.

At the bottom of the epidermis lies the basal layer, where cell division generates new cells every day. Because the rate of multiplication depends on the body's available energy, the process usually occurs during the four hours after midnight when the body's metabolism has slowed down. In a cycle lasting about 27 days, the new cells move upward through the epidermis, gradually changing from the soft, columnar cells of the basal layer from which they are eventually shed. The cells are attached to each other by plaques called desmosomes and ascend toward the surface of the skin in a continuous impermeable layer.

The horny corneal layer, composed of dead epidermal cells averaging 20 cells deep, creates a durable, protective barrier for the layers beneath it. A thin structure called the Reins barrier is resistant to salt and water, further protecting living cells from the damaging effects of excessive dehydration.

Skin

EACH SQUARE INCH
Each square inch of human skin consists of 20 ft. (6 m) of blood vessels.

THE MULTILAYERED SKIN

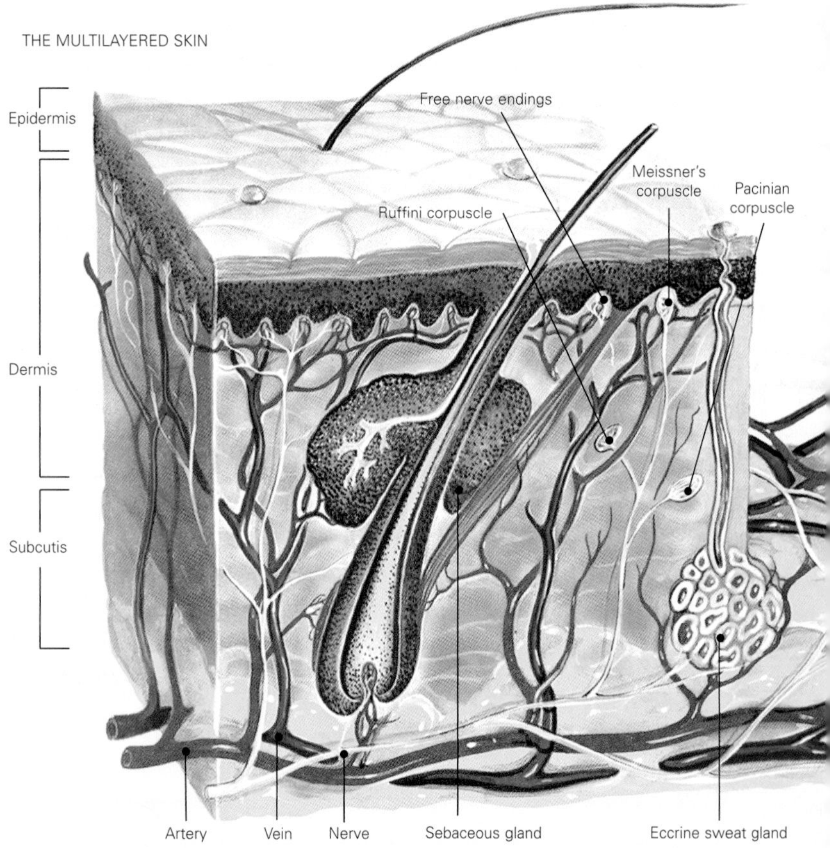
Epidermis
Dermis
Subcutis
Free nerve endings
Meissner's corpuscle
Pacinian corpuscle
Ruffini corpuscle
Artery
Vein
Nerve
Sebaceous gland
Eccrine sweat gland

Skin

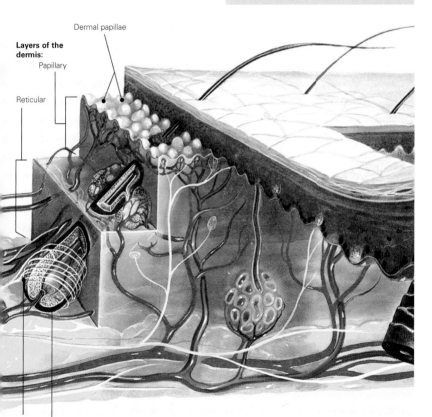

Dermal papillae

Layers of the dermis:

Papillary

Reticular

Hair end organ

Hair follicle

Skin

A Fibrous and Fatty Fabric

The dermis, or true skin, is thick, sturdy and subtle. Rich in nerves and blood vessels, as well as sebaceous and sweat glands, it shields and repairs injured tissue. Also known as the corium, the dermis, like the epidermis, is thickest on the palms and soles. This layer consists primarily of collagen, which originates from cells called fibroblasts, and is one of the strongest proteins found in nature. It gives skin durability and resilience.

Certain fibroblasts produce another protein, elastin, a relative of collagen but more easily stretched. Elastin fibers, while especially abundant in the scalp and face, are found everywhere. Collagen and elastin permit the skin to stretch easily, yet quickly regain its shape. They enable the skin to bend and fold with the body's unceasing motions. Collagen also builds a scar tissue to heal skin damaged by cuts and abrasions.

The dermal papillae carry nerve endings and capillaries to the living layers of the epidermis. When capillaries rupture, blood leaks into surrounding tissue, giving rise to the black-and-blue discoloration of a bruise. Papillae also equip the epidermis with lymph capillaries which carry away cellular wastes and help dispatch toxins and dangerous organisms. Visible evidence of the dermal papillae also shows up in skin lines, most notably in the ridges of fingerprints.

Deeper in the dermis lies the reticular layer, a dense but elastic fabric of the collagen fibers. Here, junctions of blood vessels control blood circulation through the skin and so help regulate body temperature and blood pressure. Also packed into this layer are sweat glands, hair follicles, and oil-producing sebaceous glands.

PROTECTIVE WRAPPING

Along with a layer of fat underneath, skin insulates you against all kinds of bumps, bangs, and wear and tear, keeping germs and water out and your body's fluids and salts in.

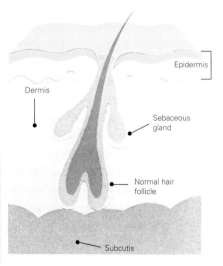

Epidermis

Dermis

Sebaceous gland

Normal hair follicle

Subcutis

Skin

The subcutis, joined to the bottom of the dermis, is the deepest layer of the skin. Just as keratinocytes synthesize keratin for the epidermis and fibroblasts furnish collagen for the dermis, so lipocytes make lipids for the subcutaneous tissue. This fatty layer cushions muscles, bones and internal organs against shocks, and acts as an insulator and source of energy during lean times.

LIVE SKIN
As skin is alive, it is made of many thin sheets of layers of flat, stacked cells in which you'll find nerves, blood vessels, hair follicles, glands, and sensory receptors.

BELOW One of the commonest skin blemishes—and for teenagers one of the most distressing—is the comedo, or blackhead. Large numbers of blackheads occurring together result in acne. The chief culprit is sebum, the mixture of waxes and fatty acids produced by the skin's sebaceous glands. During puberty, increased hormonal activity encourages these glands to produce more sebum. Dead skin cells accumulate in the minute pit round the base of a hair, mix with sebum, and form a plug that stops further sebum from escaping. Enzymes from bacteria in neighboring sebaceous glands then get to work, forming irritant acid substances and pus. With nowhere else to go, the pus bursts the hair follicle and seeps into the dermal tissue. The skin becomes inflamed, and a white-headed pimple forms. Deeper skin lesions may form pus-filled cysts.

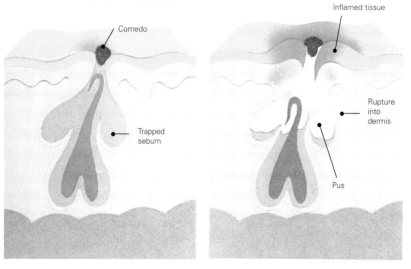

Comedo

Trapped sebum

Inflamed tissue

Rupture into dermis

Pus

Skin

THE ALL-PURPOSE OIL

Like the hard cells of the corneal layer, hair grows from the living tissue of the epidermis. The hair root, however, reaches deep into the subcutaneous tissue and rises from a special structure, a follicle, which is a thin sac of epidermal tissue with a bulb at its base.

Adjacent to the hair follicle, and connected to it by a short duct, are two sebaceous glands that provide oil for the hair and outer skin. These glands produce sebum, a mixture of waxes, fatty acids, cholesterol, and debris of dead cells. Unique to mammals, sebum coats both hair and fur with a waterproof shield that helps insulate the body from the

RIGHT In winter's cold or summer's heat, the skin reacts to the body's needs. In the cold (left), tiny arrector muscles pull hairs erect to trap an insulating layer of air next to the skin. The anatomoses between capillary veins and arteries become narrowed to reduce blood flow to the cool skin surface. Heat or vigorous exercise causes opposite effects (right). The arrector muscles relax, and sweat glands pour moisture on to the skin, where its evaporation produces cooling. Anastomoses widen, so that blood reaches the skin surface to shed heat.

WRINKLES
A baby's skin is young and fresh, and has plenty of elastin. As it becomes older, it is affected by wind, sun, and rain. The number of microscopic elastin fibers also decreases. This means that the skin becomes less flexible, making lines and creases appear. This process is speeded by exposure, especially to the sun.

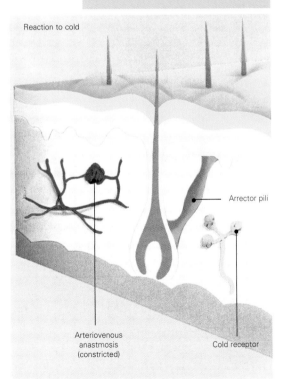

Reaction to cold

Arrector pili

Arteriovenous anastmosis (constricted)

Cold receptor

Skin

rain and cold. Since humans have lost all but a few patches of hair, and have clothed and sheltered themselves against the elements, they no longer need much sebum for insulation. But their glands still produce it in abundance.

Aside from enhancing the insulating properties of hair, sebum serves several other functions. By coating the dead

NAILS AND HAIR
Nails grow at a rate of about 3mm a month. Like hair, they are made of dead cells and have no feeling except at the root, where new nail grows—that's why cutting them doesn't hurt. The root is the bottom of the nail, under the cuticle. Nail and hair are made of tough material called keratin.

keratin cells of the corneal layer and the hair, sebum retains moisture, keeping hair glossy and skin pliable. Sebum also contains a chemical, which when catalyzed by the ultraviolet rays of the sun, becomes vitamin D. In addition sebum can kill certain forms of harmful bacteria.

Earwax, dandruff, and the crusty substance that collects around the eyes during sleep are all forms of sebum. Indeed, the face and scalp, where there are proportionately large numbers of sebaceous glands, are particularly rich in it. Sebaceous glands coat the skin continuously, more copiously in men than in women because of the effect of the male sex hormone testosterone.

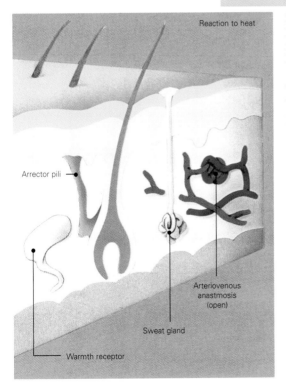

Reaction to heat

Arrector pili

Arteriovenous anastmosis (open)

Sweat gland

Warmth receptor

Skin

Occasionally, the glands supply too much sebum. This condition, called seborrhea, gives the hair and skin a greasy sheen. Even with a normal rate of sebum production and regular washing, the sebaceous glands and hair follicles of the face, neck, and upper back create conditions favorable to acne, perhaps the most common disorder of the skin.

TWO REASONS TO SWEAT

The body produces different kinds of sweat from two types of gland, the apocrine and eccrine. Apocrine glands become active during adolescence and are therefore considered a sexual characteristic. They respond not to heat but to the excitement of fear, anger, sexual arousal, and other strong feelings. The apocrine glands respond immediately to these emotions, secreting a small amount of cloudy fluid, squeezed out by a muscular sheath. Like the sebaceous glands, they open into the hair follicle to gain access to the skin.

Apocrine glands are found only in the armpits, ear canals, and around the nipples, and genitals. They may be the vestigial remains of a once prominent scent system that also played a role in social behavior, perhaps producing a sexual stimulant, or a territorial marker.

The sweat that dampens the brow comes from eccrine glands. Spread over the body they number between two and three million, and secrete a greater profusion of sweat than apocrine glands. The output of eccrine glands can total as much as three gallons a day in hot weather.

As a thermoregulator, the eccrine system reacts to stimulation from the hypothalamus in the brain by moistening the skin. The sweat, almost totally composed of water, evaporates and draws excess heat out of the body, thereby maintaining a constant internal temperature. Certain eccrine glands respond to other stimuli. Glands situated in the forehead, underarms, palms, and soles, for instance, work at times of psychological stress, independent of heat or muscular exertion.

THE SKIN'S RESIDENTS

Human skin appears lifeless to the naked eye. But the number of living organisms on a person's skin roughly equals the number of people on the planet. The fauna and flora of the skin are permanent residents. Harmful organisms attract the most attention, but the overwhelming majority of the skin's residents are harmless.

DUST MITES

Right now, there are over a million dust mites, microscopic critters invisible to the human eye, on your mattress and pillow, chomping on dead skin cells that fell off you last night!

Skin

The skin varies from one region of the body to another. Different populations of bacteria and yeasts have adapted to specific environments. The dry expanse of the forearm, the dense tangle of the scalp, and the oily surface of the nose all harbor particular species.

One of the largest residents of the skin lives in the hair follicles of the eyelashes, nose, chin, and scalp of most adults. A narrow, wormlike mite, *Demodex folliculorum*, lives most of its life in the hair follicle and lays its eggs in sebaceous glands. The young mites molt twice in the follicle, then journey across the skin at night in search of another follicle to colonize.

THE BACTERIAL FRONTLINE

The largest community of the skin's residents are bacteria, which are acquired at birth. In natural deliveries, babies pick up bacteria when passing through the mother's vagina. In the days following birth, the bacteria population expands rapidly. Within a day, bacteria in the armpit will number about 36,000 per square inch. In another four days, their number reaches 144,000 and by the ninth day the population levels off at around 490,000. Bacteria spread from person to person both by contact and by constant shedding of skin cells.

In addition to the skirmishing lines of resident bacteria, the skin relies on several other defensive strategies to ward off harmful microorganisms. Daily skin loss prevents many would-be colonists from gaining a foothold on the skin. The skin also presents an acidic mantle that deters certain types of bacteria. When the skin's bacteria breaks down sebum, fatty acids that increase the skin's acidity are produced. Bacteria thrive in moist areas, so the dryness of the skin probably accounts for most of its resistance to bacterial infections. When the skin becomes too moist, its resistance decreases and harmful bacteria break through normal defenses.

KEEPING THE TEMPERATURE NORMAL

The skin plays a prominent role in maintaining the body's temperature. The job is a constant one, requiring small adjustments to ensure that the body's core temperature does not stray from the narrow range at which the organs function efficiently. This temperature—which differs slightly from person to person—averages 98.6°F. Called the set point, it is analogous to the setting of a

FINGERPRINTS

Your fingerprints are quite unlike anyone else's. The ridges of skin on the fingertips form a pattern, and everyone's pattern is different. People's palm prints are also different. One way of doing this is to make ink prints of your hands.

Skin

thermostat. The system is so delicately balanced that if the core temperature varies by 1.5 degrees, the body's metabolism is altered by about 20 percent.

When body heat shifts slightly from the set point, the heating or cooling mechanisms quickly restore the proper temperature. The skin achieves this regulation largely by controlling the amount of heat loss. To do so, it works in concert with the hypothalamus, a cluster of nerve cells at the center of the brain. Specialized regions of the hypothalamus contain heat-sensitive and cold-sensitive cells which respond to changes in blood temperature by increasing the number of nerve impulses they transmit. On receiving these commands, the skin hastens to make the appropriate adjustments in its domain.

Even in stable surroundings, this temperature-regulating system continues to function all the time, for although the body constantly produces heat, it also constantly loses heat. At a moderate room temperature, the body loses most heat through what scientists call radiation— rays of heat that emanate in all directions. Other objects in the room constantly radiate heat back to the body, but a body warmer than its surroundings always loses more heat than it gains.

RESPONDING TO THE COLD AND HEAT

Most cold detection occurs on the body's periphery, which contains abundant specialized nerve endings called thermoreceptors. The skin has far more receptors to detect coldness than receptors to detect warmth. The number of cold receptors also varies from one region of the body to the next. The skin on the lips contains about 20 times more cold receptors than the skin on the chest or legs.

Cold receptors relay their signals via the spinal cord to the hypothalamus, which sends impulses to structures in the skin called arteriovenous anastomoses. These are specialized connections between veins and arteries that intermittently substitute for capillaries throughout the circulatory system when blood flow needs to be rerouted. They are particularly abundant in the hands, feet, eyelids, nose, and lips. The middle portion of an anastomosis is a thick, muscular wall that contracts, restricting blood flow to the extremities and thus reducing heat loss. This is why our lips and fingernails turn slightly blue when we are cold.

STRAIGHT OR CURLY

Skin feels smooth, but under a microscope it looks like a jagged mountain range with huge pits of sprouting hair. These pits are called follicles, and they make hair straight or curly. Straight hair grows from a round follicle, wavy hair from an oval-shaped one, and very curly hair grows from a flat follicle.

Skin

With restricted blood flow, tiny lumps, resulting from erection of the hair shafts, cause goose bumps to appear on the skin. This action, or piloerection, is largely useless in humans. In animals, it makes fur stand on end and helps create an extra layer of warmth by trapping warm air next to the body.

The body reacts to excess heat by taking measures to encourage heat loss. Sweating begins and blood vessels in the skin dilate to allow more blood to reach the surface. The blood dissipates more body heat, accounting for the flushed, rosy glow of the skin during vigorous exercise. This same mechanism causes blushing. The layer of sweat that builds up on the skin's surface helps cool it by evaporating.

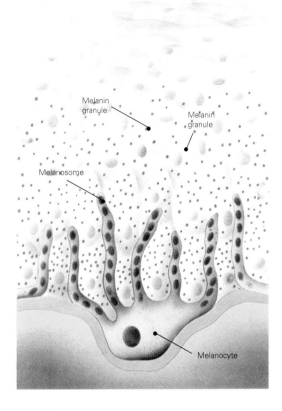

Melanin granule

Melanin granule

Melanosome

Melanocyte

GOOSE BUMPS

Each hair has a tiny erector muscle. When you are cold, these muscles contract to make the hairs stand up, trapping warm air between them and giving you goose bumps.

ABOVE A golden suntan results from the activities of pigment cells in the skin. Called melanocytes, they are triggered into action by ultraviolet radiation from the sun or a sunlamp. They increase their output of the dark pigment melanin and inject it into nearby epidermal cells. The cells undergo their normal migration toward the skin's surface, so that in four or five days they arrive there, tanning the skin and forming a filter to block some of the sun's rays and preventing possible damage.

Skin

THE HAIRY HUMAN

Like all mammals, human beings are covered in hair. But unlike the thick fur coats of four-footed members of the group, our "fur" consists, for the most part, of fine hairs that grow few and far between. The only thick growths of hair are in the pubic region, under the arms, on the head, and on the faces of men—although children have no pubic, axillary, or facial hair and many men lose much of their head hair after middle age.

Every square inch of the body—except the lips, palms of the hands, and soles of the feet—has hair growing on it. Some is so fine as to be almost invisible, although it reveals its presence as the site of every goose bump when the body is cold. Each hair grows like a stem of grass from a bulblike follicle, which tunnels through the surface of the skin. The root forms inside the follicle and the shaft grows out of the skin. A combination of black, brown, and yellow pigments give the hair its color. Cells in the outer layer, or cuticle, of the shaft overlap like a fish's scales. A fine nerve fiber encircles the follicle, which also has a cluster of sebaceous glands draining into it. They coat the hair with oily sebum, which lubricates and protects it.

BEARDS
Beards are the fastest-growing hairs on the human body. If the average man never trimmed his beard, it would grow to nearly 30 ft. (9 m) long in his lifetime.

THE PIGMENTATION OF THE SKIN

The wide variety of human skin color is a direct measure of the diverse climates to which man has adapted. All skin color stems from the same substance, the pigment melanin. Two forms of melanin color the skin, hair, and eyes of humans. The major pigment, eumelanin, produces shades of brown and black; the other, phaeomelanin, is the pigment of red hair.

Melanocytes are spider-shaped, with long irregular arms that reach out from the cell body. The arms of each melanocyte link it with about ten surrounding cells. Melanocytes inject pigment granules, melanosomes, into these neighboring cells, thus spreading pigment across the skin. Freckles, a result of the pigment's gathering in clusters, can occur in all skin colors but are most prominent in light skin.

Regardless of skin color, all human beings have approximately the same number of melanocytes in the skin—around 1 percent of all skin cells. Differences in skin color are due to the amount of melanin the cells produce, which in turn depends on how much ultraviolet radiation they must absorb. Melanin absorbs ultraviolet rays and converts them into harmless infrared rays. The more melanin there is in the body the more efficient this absorption becomes, and the better the protection afforded.

Skin

The reason why everybody gets a tan or sunburn when exposed to the Sun for long perios of time is your skin's way of telling you it is protecting itself from the Sun's harmful rays. The table below shows different reactions from different skin types.

SKIN TYPES AND THEIR SKIN REACTIONS TO THE SUN

SKIN TYPES	EXAMPLES OF THE SKIN TYPE	REACTION TO THE SUN
TYPE 1	Fair skinned people, blue-eyed people, sometimes brown-eyed people, freckles, reddish, or blonde-haired people.	Burns all the time, painful at times; rarely tan; end up peeling; very sensitive to UV rays.
TYPE 2	People of fair skin, blonde, brown, red hair, and brown, hazel, or blue eyes.	Always burn, painful at times; sometimes tan, but not often; could peel; sensitive to UV rays.
TYPE 3	Caucasian.	moderately burns, will tan gradually; sensitive to UV rays.
TYPE 4	People with olive, light brown, or white skin, dark brown hair, and dark eyes; moderately sensitive to UV rays.	Does not burn often; tans easily and reacts quickly making the skin dark.
TYPE 5	Brown-skinned people like people of Asian, Hispanic, or Indian descent.	Rarely burns; tans easily; reacts quickly making the skin dark; can minimally withstand UV rays.
TYPE 6	African-Americans.	Never burns, but tans a lot; skin reacts quickly making the skin dark; can withstand UV rays.

BRAIN

The human brain is what sets us apart from other animals. It gives us the ability to reason, to communicate with others, to learn, and to remember. These skills have made human beings the dominant species on planet Earth, able to tame and manipulate the environment, build ships and cities, travel to the moon, and eventually reach other planets and perhaps, one day, the stars.

The brain is also the seat of the "human" qualities of love, compassion, mercy, and forgiveness. Painting, poetry, music, and drama—all of humankind's artistic achievements emanate from an irregular blob of jelly that makes up about two percent of the average person's body weight.

A source of mystery during the entire span of human history, the brain has recently yielded up many of its secrets—but not all. For many years to come, the brain will continue to puzzle scientist and philosopher alike.

MANAGEMENT FROM THE TOP

Ballooning from the top of the spinal column like some weird, science fiction flower, the brain is the body's central computer which controls every thought and most movement. Information from all parts of the body is carried by sensory nerves to the brain, where it is integrated with direct input from the external senses. After the correct decision has been made, instructions for action are sent down motor nerves to the muscles. Not all of the action is conscious and voluntary; the brain also controls the "automatic" processes of the body, such as breathing, heart rate, and digestion.

The adult brain weighs about 3 lbs. (1.3 kg), has a highly wrinkled appearance, and contains some ten billion nerve cells. By far the largest part is the ovoid mass of the cerebrum, where thinking takes place. Hanging beneath the back of the cerebrum is the cerebellum (the "little brain"), and jutting out from the middle is the brainstem, which merges into the spinal cord.

TWO SIDED

Each hemisphere, or half of the cerebral cortex, works slightly differently, carrying out the major part of a process, such as creating a work of art or thinking through a tricky problem.

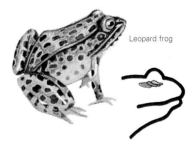

Leopard frog

Brain

The brain and spinal cord consist largely of gray and white matter. Gray matter is made up of nerve cell bodies, or neurons; white matter is nerve fibers, or axons, in their insulating myelin sheaths. In the spinal cord the gray matter is contained within the white matter, in the brain the gray matter is on the outside.

CARE AND PROTECTION

The tissues of the brain are protected from external knocks and damage by the strong, bony dome of the skull. Inside the cranial cavity, further protection is provided by a cushioning layer of soothing cerebrospinal fluid, a liquid secreted in the ventricles, or cavities, deep within the brain.

The brain is at once delicate and resilient. Large parts of it can be removed with no apparent effect, yet slight damage to other parts can result in total disruption of body functions and death. Much of our basic knowledge of the brain's operation comes from observing the effects of accidental damage or tumors. Electrical stimulation during surgery has also provided many clues. However, even the finest microscopic probes seem clumsy when applied to tissue in which a piece the size of a pinhead may contain up to five million cells.

The greatest source of information about the human brain is through experiments on animals. Because all mammals share the same basic brain structure, results of experiments on cats or monkeys can be applied to humans. However, scientists must be aware that it is the greater development of the brain, especially the forebrain, that makes us human, and not an animal.

BRAIN SIZE
Human beings do not have the largest brains—that distinction must go to the elephants and whales. But the human brain is the largest in proportion to the size of the body that holds it, and weighs up to three pounds. It also differs in the relative proportions of its chief parts, the buff-colored cerebrum and the evolutionary older hindbrain.

Fish, striped bass

Brain

UP THE CHAIN OF COMMAND

The division of the brain into three basic parts—hindbrain, midbrain, and forebrain—is based on their relative positions ascending from the top of the spinal column. An alternative concept is based, upon three stages of the evolution of life on planet Earth. The first is reptilian, followed by paleomammalian (the very first mammals), then neomammalian (the later mammals, including humans). Although there is

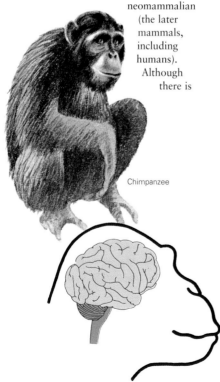

Chimpanzee

considerable overlap between the two concepts, the one based on location provides a more accessible description of the structures that make up the brain.

At the top of the spinal cord, the hindbrain was the first part of the brain to evolve. Comprising the cerebellum and most of the brainstem, it is sometimes referred to as the primitive brain.

In the region where the spinal cord gradually merges into the brainstem lies the medulla oblongata. This fibrous first inch of brainstem is the body's reflex center, the evolutionary core, and primitive site of survival in human or dinosaur. Here are the small clusters of gray matter that control respiration, heartbeat, blood pressure, and gastric movements. From the medulla also flash the involuntary commands to swallow, sneeze, or cough.

MORE COMPLEX
The human brain is the most complex of any animal on earth. This is not because they are the biggest, but human brains are larger and heavier in comparison to the human body.

Brain

Source of human thought and creativity, the cerebrum occupies two-thirds of our comparatively massive brain. In most other animals, the cerebrum takes up much less of the brain's space. Only higher mammals, such as those represented on these few pages by the cat (overleaf) and the chimpanzee (opposite), have cerebrums of anything like human complexity.

BILLIONS OF CELLS
Although our brain is wrinkled, soft, and a little wet, and doesn't look like much, it is made up of more than 10 billion nerve cells and over 50 billion other cells and weighs less than 3 lbs. (1.3 kg).

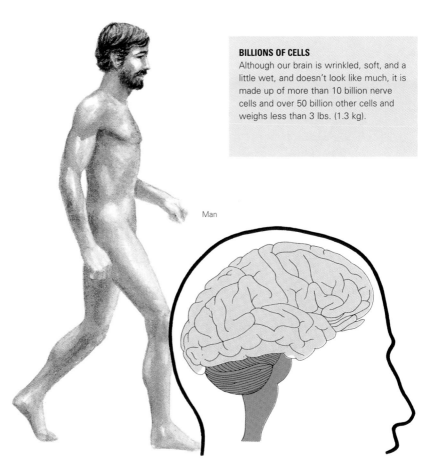

Man

Brain

NERVOUS DISPOSITION

As well as being the body's "automatic pilot," the medulla is also a major traffic intersection. Two thick bundles of nerve fibers, known as pyramids, run down the medulla linking the brain to the muscles of the limbs. At a point known as the decusation (meaning "X-shape") of the pyramids, most of the fibers cross over so that the left side of the brain controls the right side of the body, and vice versa.

Branching out from the medulla and adjacent areas above the decusation are the 12 pairs of cranial nerves, which serve the sensory and motor needs of the head, neck, chest, and abdomen. Counting from front to back, and top to bottom, they are known by both name and roman numeral: olfactory (I), optic (II), oculomotor (III), trochlear (IV), trigeminal (V), abducens (VI), facial (VII), acoustic (VIII), glossopharyngeal (IX), vagus (X), spinal accessory (XI), and hypoglossal (XII). To remember their initials in correct order, anatomy students recite, "on old Olympus' tiny tops, a Fin and German viewed some hops."

FOLDS AND CRAGS
Underneath your scalp, sits your brain which, as it has grown, has continued to fold in on itself and develop deeper and deeper folds and crags. Spread out it would be the size of a pillowcase.

Pigeon

Cat

Brain

THE VIEW FROM THE BRIDGE

Arching over the medulla is the pons (named from the Latin for "bridge"), an inch-wide span of white matter which, in both function and appearance, links the cerebrum of the forebrain with the cerebellum of the hindbrain. From the pons itself extend a third of the cranial nerves, including the largest, the trigeminal. As its name suggests, this divides into three, with fibers fanning out to the jaws, face, and scalp. Also located in the pons is a small piece of gray matter that controls the glands that produce saliva and tears.

Scattered throughout the pons and medulla, and extending up into the midbrain, are areas of closely interwoven gray and white matter called the reticular formations. These are the brain's chief watchdogs. Operating continuously during periods of wakefulness, the reticular formations are in essence the physical basis of consciousness, sending out the impulses that keep you awake and alert. When these impulses slacken, you fall asleep; if the formations are damaged, then prolonged unconsciousness, or coma, may result. And if they do not repair, coma persists.

RIGHT Internal shock absorption is provided by the pools of cerebrospinal fluid which bathe the brain and cushion its membranes. The plasmalike liquid is produced in choroid tissue and fills the ventricles, four cavities within the cranium (shown in blue). From the lowest fourth ventricle the fluid circulates down the spinal cord.

Lateral ventricles

Cerebrum

Third ventricle

Temporal horn

Occipital horn

Fourth ventricle

Cerebellum

CONTROLLING OPPOSITE SIDES
Your brain is divided into two sides. The left side of your brain controls the right side of your body, while the right side of your brain controls the left side of your body.

Brain

ALARMING DEVELOPMENT

Serving as the gatekeepers of consciousness, the reticular formations sort out the hundred million nerve impulses that assault the brain every second, separating the important from the inconsequential. The brain does not need a constant report of exactly how your shoes feel upon your feet; therefore, only about a hundred impulses per second are permitted through to the regions above the brainstem. Of these, the conscious mind heeds but a few. While you may be vaguely aware of nearby sights and sounds, concentration is limited to one at a time.

Together with their associated nerve pathways, characterized by short fibers enabling impulses to travel at high speed, the reticular formations also provide the body's alarm system: the reticular activating system (RAS). A nerve impulse requiring urgent attention, such as the smell of smoke, is detected by the RAS and sent straight to the cerebrum for action, temporarily overriding all other messages.

OPERATOR

Like a telephone operator who answers incoming calls and directs them to where they need to go, similarly, your brain acts as an operator by sending messages from all over the body to their proper destination.

This mechanism also works in reverse. The conscious mind can instruct the RAS to block out everything else, and focus attention on one particular action that requires the highest skill and concentration. Without the RAS, a tennis champion could not ignore the murmurs from the crowd while serving for the match at Wimbledon.

The RAS also regulates a process known as habituation. An unusual sensation, such as the sound of a jack hammer in the street outside, immediately sets off a RAS alert. As soon as the sound has been recognized and the novelty has worn off, the brain instructs the RAS to halt its alerting action.

SMOOTH PERFORMER

Attached to the pons, and bulging out behind it, the cerebellum makes up about one-eighth of the brain's total mass. Its twin lobes are extremely wrinkled, and 85 percent of the surface area is hidden in the numerous deep folds. Working closely with the organs of balance in the inner ear, your cerebellum controls posture and governs your every movement. Apparently initiating nothing itself, it monitors impulses from motor centers in the brain and from nerve endings in the muscles.

Modifying and coordinating commands to swing and sway, the cerebellum grooves a golfer's tee shot, smoothes a dancer's footwork, and lets you lift a glass of water to your lips

Brain

without spilling a drop. Learning a sequence of actions, like riding a bicycle or touch typing, takes both time and trouble, and many parts of the brain are involved. Once the sequence has been learned, however, the cerebellum is able to take over and ensure that it is repeated without conscious effort. There is evidence that the cerebellum plays a part in a person's emotional development, modulating sensations of pleasure and anger.

NEWBORNS
During the first year of life a newborn baby's brain will grow almost three times larger.

BELOW Conventionally identified with Roman numerals, the 12 pairs of cranial nerves branch out like power cables from the top of the brainstem. On the diagram, purple represents sensory nerves, which carry messages to the brain from sense organs and other receptors around the body. Orange is the color given to motor nerves, which transmit instructions from the brain to muscles in the face. An exception is the cranial nerve X, the vagus, whose motor fibers connect with the heart and digestive system.

Brain

WELL CONNECTED

A section through the two lobes of the cerebellum reveals matching leaflike patterns caused by the branching network of white nerve fibers. This prompted early anatomists to name it "arbor vitae," or tree of life. The exterior covering, or cortex, of gray matter is made up of three distinct layers. The middle layer consists of Purkinje cells, which are among the largest and most complex nerve cells in the human body. Each Purkinje cell can interconnect to up to a hundred thousand others, and no other cells in the brain have so many potential connections; such is the organization required for precise muscle control.

Between the cerebellar lobes lies a small structure known, for its wormlike shape, as the vermis. From it protrude three pairs of nerve fiber bundles known as peduncles. The lower two pairs, the inferior and middle peduncles, connect to the medulla oblongata and pons, respectively. These carry incoming, or afferent, impulses to the cerebellum. The superior peduncles above carry outgoing, or efferent, impulses to the red nucleus of the midbrain and thence via the thalamus to the cerebrum. Despite its crucial role in regulating our movements, the cerebellum does not exercise any direct control.

RELIC OF THE PAST

Above the bridge of the pons lies the midbrain, the last inch of the brainstem. Before the evolution of the cerebrum, the midbrain was responsible for the higher functions of the brain, such as processing the senses of sight and sound. These have now migrated to the forebrain, and in consequence the structure of the midbrain is much simpler than either the forebrain or hindbrain. Some evidence of its former eminence is still visible in the four bumps of the colliculi, which serve as relay stations for nerve impulses from the eyes and ears.

Two further peduncles rise to connect the brainstem to the hemispheres of the cerebrum. Each contain a dark core of cellular matter stained with the pigment melanin. This is the substantia nigra, motherlode of the biochemical agent dopamine, which guards against muscle rigidity and tremor. Down the center of the midbrain runs the cerebral canal, which carries cerebrospinal fluid by now drained of nutrients and loaded with metabolic waste, down into the space surrounding the spinal cord where it is absorbed into the blood.

THREE PARTS
The brain is made up of three parts: the forebrain, the brainstem, and the hindbrain. Of these three parts, the most complex is the forebrain which gives us the ability to "feel," learn, and remember.

Brain

PRIMITIVE PASSION

The largest and most recently evolved part of the forebrain is the cerebrum, which is divided into the two cerebral hemispheres. In fact these are quarter spheres—the whole cerebrum makes a hemisphere—but the name has stuck. Within each cerebral hemisphere is a wishbone-shaped cavity where cerebrospinal fluid is produced, replenishing the supply three times per day. Known as the lateral ventricles, these drain into the centrally located third ventricle, down through the small fourth ventricle, and into the cerebral canal.

A complex of small structures, notably the thalamus and hypothalamus, cluster around the third ventricle and loop over the top of the brainstem; they make up the remainder of the forebrain. This is sometimes known as the "interbrain." Together with parts of the cerebrum and the pathways associated with humans' almost redundant sense of smell, this "interbrain" forms the limbic system—the most primitive part of the forebrain and the seat of human passions and basic drives.

THE BRAINSTEM
The brainstem consists of the midbrain and hindbrain. Just as the name suggests, the brainstem resembles the stem of a branch.

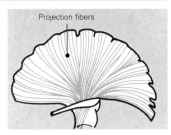
Projection fibers

PROJECTION FIBERS Forming the *corona radiata*, these fibers fan out from the brainstem. They relay impulses to and from the cortex.

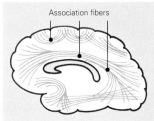
Association fibers

ASSOCIATION FIBERS Looping strands link different sections of the same hemisphere. This web subtly modulates the cerebral cortex.

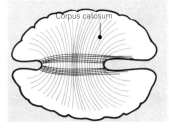
Corpus callosum

CORPUS CALLOSUM This thicket of fibers joins the hemispheres, permitting the two sides of the brain to communicate with each other.

ABOVE RIGHT Cross connections within the brain are provided by various bundles of nerve fibers. A fan of projection fibers connects the brainstem with the cortex (top). Within each hemisphere association fibers, shown in the sideways section (middle), interconnect various parts of the cortex. To prevent the two hemispheres working in isolation, they are joined by the fibers of the corpus callosum, depicted in the horizontal section (bottom).

Brain

BALANCING ACT

Here at the very core of the brain is one of its most subtle and little-understood regions. Hidden beneath the enveloping folds of the cerebral hemispheres, the limbic system has been termed the "emotional brain," striving to maintain an equilibrium between fear and desire. It also attends to everyday wants and needs. The nerve cell masses in the system are interwoven with myriad pathways that convey not only the extremes of terror and ecstasy, but also more mundane electrochemical messages such as when to sweat and when to shiver, what to remember, and what to forget. The limbic system is the body's balancing act on the tightrope of survival.

While the full workings of the system remain obscure, the basic role of the major components is understood. The thalami, twin eggs of gray matter perched above the top of the brainstem, are relay stations connected to a very large number of information channels that bring input from the main sensory systems, the cerebral hemispheres, the cerebellum, and the reticular formations in the brainstem. Within the thalamus this mixed input is integrated and information patterns of

> **SEPARATION**
> The left and right regions of the cerebral cortex are separated by a thick band of tissue called the corpus callosum.

greater complexity are sent back. The thalamus appears to be concerned with emotional shading, subjective feeling states, and awareness of identity and self.

THE LITTLE DICTATOR

The size of a thumb nail and with one of the richest blood supplies in the body, the hypothalamus is the body's great regulator, maintaining internal temperature and fluid balance. Tiny receptors measure the amounts of glucose and salt in the blood, generating feelings of hunger or thirst when levels become low. Through its neighbor, the pea-sized pituitary gland, the hypothalamus organizes the release of hormones which affect growth and sexual behavior. It also initiates the "fight or flight" response to stress, sending chemical messages via the pituitary to the adrenal glands above the kidneys. Chemical control alone is not enough for the brain's little dictator, and cell clusters specializing in involuntary muscle control crackle with electrical energy as they fire off signals.

Although signs of crude rage and feelings of raw pleasure can be obtained by experimentally stimulating parts of the hypothalamus, more complex emotions, and "instinctive" behavior patterns such as finding a mate or rearing young, reside in the interaction of the limbic system as a whole. Above the hypothalamus, the amygdala is thought to modify rage and aggression to suit rapidly changing circumstances. To the rear, projecting

Brain

from the roof of the third ventricle, is the pea-sized pineal gland. Once believed to be a vestigial third eye, it is probably a light-sensitive biological clock, which regulates sex-gland activity. Above the thalami a fibrous web, the fornix, terminates at the bulbous hippocampal formations. The hippo-campus converts information from short-term memory to long-term memory, and constantly

CORTEX
The word "cortex" comes from the Latin word for "bark" (of a tree). This is because the "cortex" is a sheet of tissue that makes up the outer layer of the brain, varying from 2 to 6 mm thick.

compares signals from the senses with stored experience.

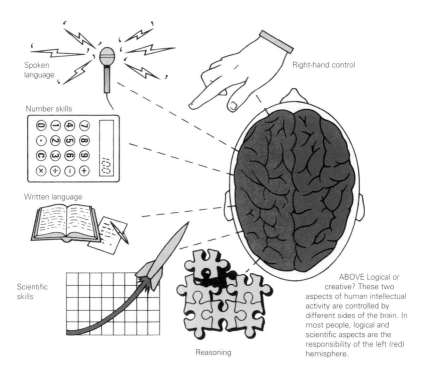

Spoken language

Right-hand control

Number skills

Written language

Scientific skills

Reasoning

ABOVE Logical or creative? These two aspects of human intellectual activity are controlled by different sides of the brain. In most people, logical and scientific aspects are the responsibility of the left (red) hemisphere.

Brain

CONVOLUTED THINKING

The cerebrum, arching beneath the curvature of the skull, is the seat of reason, imagination, and creativity. Its surface is molded into a series of convolutions called gyri, separated by fissures, or sulci. The most pronounced fissure, the longitudinal cerebral fissure, separates the two cerebral hemispheres.

The convolutions of the cerebrum are not as dense and intricate as those of the cerebellum, and their arrangement is fairly constant in all brains. For ease of description, certain deeper fissures are used to divide each hemisphere into four lobes—frontal, temporal, parietal, and occipital—although this is purely convenience and these areas of the brain do not coincide exactly with the cranial bones for which they are named.

The gray matter of the cerebrum, "the little gray cells" of a famous fictional detective, is mostly contained within a thin layer covering the gyri and lining the sulci. This surface layer, the cerebral cortex, is on average about one-quarter of an inch thick. Only about 30 percent of its area is externally visible; the rest is hidden within the fissures. Underneath the cortex is a mass of white matter, the connecting fibers.

IN BETWEEN THALAMUS AND SPINAL CORD
The brainstem is a general term for the area of the brain between the thalamus and spinal cord.

TYING IT ALL TOGETHER

Hundreds of millions of microscopic threads form an array of connections between the cortex's nerve centers and distant parts of the brain. Although only a tiny fraction of the connections have been traced, their general organization is understood and they are divided into three different types.

The projection fibers squeeze into a compact band, the internal capsule, near the top of the brainstem. Both incoming and outgoing fibers then flare upward and outward to all parts of the cortex, creating a pattern that anatomists have dubbed the corona radiata. The most numerous fibers link different parts of the same hemisphere, modulating the activities of the cortex and providing built-in self-sufficiency. These are aptly named association fibers.

One of the skeins of association fibers, the cingulum, is covered with a layer of cortex to form the cingulate gyrus, a major part of the "limbic lobe" of the cerebral hemisphere. This fifth lobe is a relative newcomer in terms of the brain's known geography, identified only quite recently, and its full extent has still to be agreed. Other components are the smaller dentate gyrus and the hippocampal formations, so named because in section they resemble the shape of a sea-horse.

The "limbic lobe" is the cerebral part of the limbic system, and it is believed that here the raw energies of aggression and sex drive produced by the

Brain

hypothalamus are integrated into the rest of the brain, producing courtship and other socially acceptable behavior, and behavior necessary to the continuance of human communities.

The third category of cerebral fibers are known as commissural fibers. These are gathered into a densely packed body, the corpus callosum, located at the bottom of the longitudinal cerebral fissure. Extending beyond the brain's midline and mingling with association and projection fibers, the corpus callosum

links the cerebral hemispheres and enables the two sides of the brain to intercommunicate.

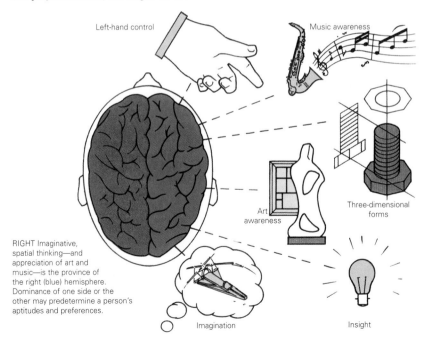

Left-hand control

Music awareness

Three-dimensional forms

Art awareness

RIGHT Imaginative, spatial thinking—and appreciation of art and music—is the province of the right (blue) hemisphere. Dominance of one side or the other may predetermine a person's aptitudes and preferences.

Imagination

Insight

Brain

CROSSING THE GREAT DIVIDE

Because of its size (it is more than 4 in. (10 cm) long) and central position, early scientists believed that the corpus callosum was essential to the brain's proper functioning. They saw it as unifying the two cerebral hemispheres into a single brain. However, in the middle of the last century it was noticed during autopsies that some people had lived apparently normal lives with half a brain, having only one cerebral hemisphere.

The mystery deepened in the 1930s when doctors discovered that epileptic seizures could be significantly moderated by surgically severing the patient's corpus callosum, and without any noticeable impairment of normal behavior. Further research, especially by a team at the California Institute of Technology led by Roger Sperry, revealed that although the two cerebral hemispheres are virtual mirror-images of each other, their mental functions are in fact very different.

The most basic distinction between the two is that the right hemisphere controls the left side of the body, and vice versa.

This at least is symmetrical, and the cross-over of nerve pathways inside the medulla oblongata is readily observable. Less apparent is the division of mental abilities—the geography of the brain merges into the geography of the mind.

TWO BRAINS IN ONE

The CalTech experimental studies of split-brain people (with the corpus callosum severed) showed that the two sides of the brain process information differently. One side specializes in symbols and logic, while the other is adept at pattern and space perception.

Left hemisphere thinking appears to be analytical (taking ideas apart), linear (one step after another), and verbal (both written and spoken). It builds sentences and solves equations. Right hemisphere thinking is synthetic (putting ideas together), holistic (grasping relationships in a single step), and imagistic (visual thinking with the "minds eye"). It listens to music and appreciates three-dimensional objects. The left side of the brain has given man science and technology; the right side is responsible for art and imagination.

Thus the proper function of the corpus callum is now understood, it exists largely to unify awareness and attention and allow the two hemispheres to share learning and memory. For his work, Sperry was awarded a Nobel prize in 1981.

CEREBELLUM
The word "cerebellum" comes from the Latin word for "little brain." The cerebellum is located behind the brain stem.

Brain

PANNING FOR GOLD

Although it is possible to chart many specific brain functions, which side of the brain contains intelligence, and where is the elusive spark of creativity to be found? The answers are not easy to come by, not least because these

highest human qualities are difficult to measure and almost impossible to define.

BELOW A section through the brain reveals its convoluted anatomy. This view shows the lobes of the cerebral cortex and vital nerve connections.

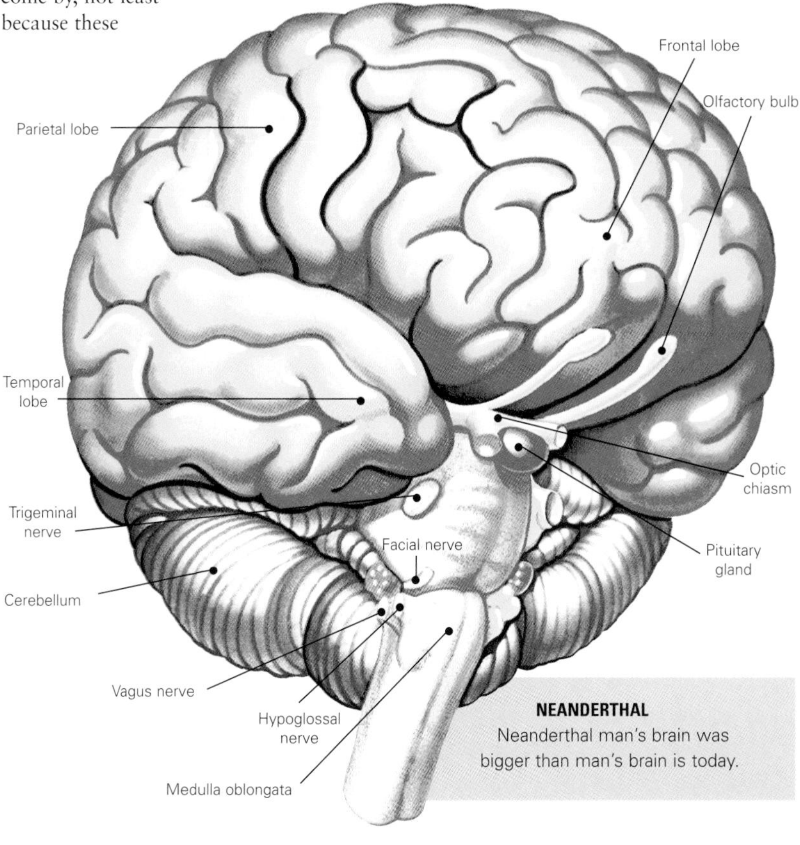

Frontal lobe

Olfactory bulb

Parietal lobe

Temporal lobe

Optic chiasm

Trigeminal nerve

Facial nerve

Pituitary gland

Cerebellum

Vagus nerve

Hypoglossal nerve

Medulla oblongata

NEANDERTHAL
Neanderthal man's brain was bigger than man's brain is today.

Brain

Traditional intelligence tests, with their sequences of numbers, missing words, and logical puzzles, are heavily geared toward left hemisphere thinking.

As such they have proved statistically reliable indicators of scholastic achievement. Outside of school, they are much less useful in predicting success in later life. How can qualities such as dealing with people or susceptibility to gambling be measured; and to what extent are these and other important factors a function of intelligence?

In a search for a better alternative some experts have urged measuring reaction times, on the basis that the quicker you think, the cleverer you are. Others have suggested using "real-life" problems such as selecting the most cost-effective route from an airline schedule. None of the tests yet devised is particularly good at measuring divergent thinking, the ability to discover new answers, which psychologists believe to be crucial to creativity.

Superficial logic suggests that creativity, the opposite of reason, ought to be lodged in the right hemisphere of the brain alongside music and art appreciation. Certainly the ability to think in images or sounds is vital to the creative process. Mozart wrote entire concertos in his head—writing them down afterward was something of an anticlimax. However, artists need more than just inspiration and creativity to survive; they must be rational and self-critical masters of their discipline. It is the interplay between the two sides of the brain that is essential for all types of productive thought. Albert Einstein observed that his gift for fantasy meant more than acquiring knowledge.

AREA OF RESPONSIBILITY

Established only by experimental study, none of the right-left separation of the brain's mental activity is visible to the anatomist's eye. Both cerebral hemispheres look the same, with irregular folds of cortex covering the whole of the surface. But the division of the hemispheres into four lobes, while partly for convenience, is not completely arbitrary because each lobe has a particular area of responsibility in the brain's organization.

The frontal lobe extends from behind the forehead up to the central sulcus that bisects the hemispheres at right angles to the main longitudinal fissure. The area directly behind the brow is confusingly known as the prefrontal lobe, and some researchers believe that powers of planning and choice are located here.

THE HYPOTHALAMUS
This is composed of several different areas and is located at the base of the brain. Although only the size of a pea, and about 1/300 of the total brain weight, the hypothalamus is responsible for some very important functions, like the control of body temperature.

Brain

Adjacent to the central sulcus is an area of cortex that controls every voluntary movement of the body—the motor cortex.

On the other side of the central sulcus, within the parietal lobe which caps the top of the cerebral cortex, is an area loaded with cells responsive to touch, heat, pain, and body position. This area is known as the sensory cortex.

LIMBIC SYSTEM
The limbic system (or the limbic areas) is a group of structures that includes the amygdala, the hippocampus, mammillary bodies, and cingulate gyrus. These areas are important for controlling the emotional response to a given situation.

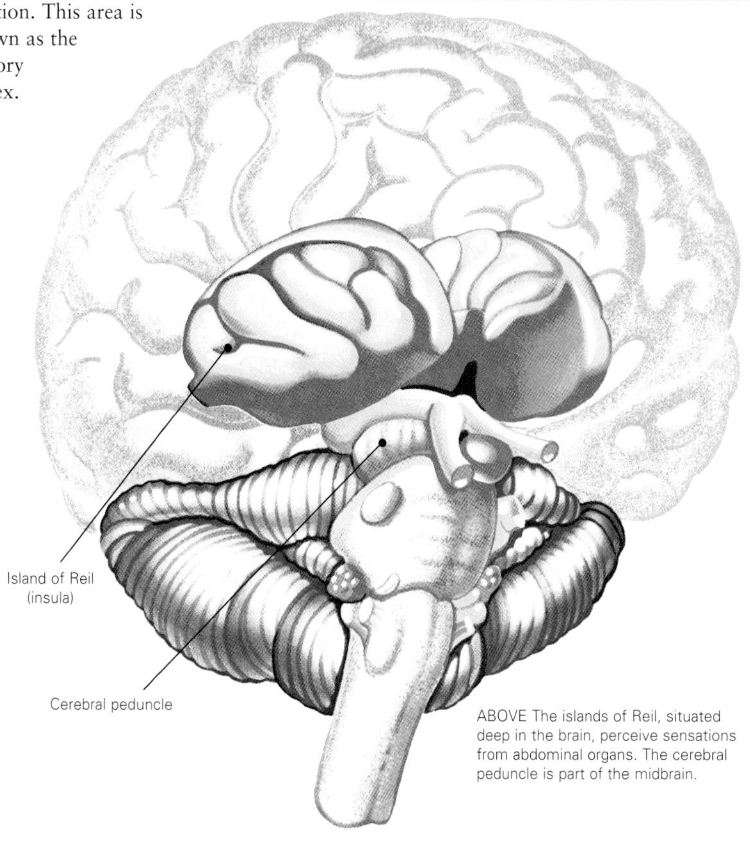

Island of Reil
(insula)

Cerebral peduncle

ABOVE The islands of Reil, situated deep in the brain, perceive sensations from abdominal organs. The cerebral peduncle is part of the midbrain.

Brain

At the back of the skull, as far from the eyes as is anatomically possible, the visual cortex occupies an area of the occipital lobe. Impulses from the retinas race down the optic nerve, half of them crossing at the junction of the optic chiasma in front of the brainstem. They then fan out through clusters of gray matter called lateral geniculate bodies before hitting the visual cortex at speeds of up to 400 ft. (121 m) per second. Not quite the speed of light, but fast enough.

In contrast, hearing is processed in the area of the brain closest to the ear. This is the temporal lobe, which is separated from the frontal lobe by the lateral sulcus at sideburn level.

These named areas of cortex have been identified because they are easily stimulated and the results can be observed. The rest of the cerebral cortex appears to lack direct connections with the central nervous system and has been dubbed the association cortex. These "silent" areas of the cortex are believed to be concerned with further elaboration of sensory information. They may also constitute the brain's memory banks.

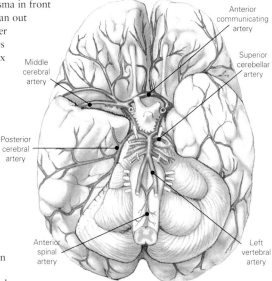

Anterior communicating artery

Superior cerebellar artery

Middle cerebral artery

Posterior cerebral artery

Anterior spinal artery

Left vertebral artery

ABOVE A constant blood supply is essential for proper functioning of the brain. A built-in fail-safe feature gives the brain the best chance of receiving an uninterrupted supply of oxygenated blood, for without oxygen brain cells die within a few minutes, and can never be replaced. The arteries that lead to the cortex branch off a closed loop. As a result, if an obstruction should occur in one place on the circuit the blood can flow around the other way to the arterial exit and on to the cortex. To remove carbon dioxide waste, produced by internal respiration of the brain's many cells, there is a network of veins on the surface of the brain that carry away the deoxygenated blood. They include internal drainage reservoirs in the form of enlarged cavities known as sinuses.

COOLING DOWN

If you are too hot, the hypothalamus detects this and sends a signal to expand the capillaries in your skin. This causes blood to be cooled faster.

Brain

PLUMBING THE DEPTHS

Deep in the floor of each lateral sulcus, completely overgrown by the rest of the brain, is the insula—a strange, vaguely pyramid-shaped structure. Also known as the islands of Reil, this has the same convoluted appearance as the cerebral hemispheres and a covering of cortex like the rest of the cerebrum. What the islands do—even what other parts of the brain they are connected to—is still very obscure. Almost the sum of our knowledge is that stimulating the insula sometimes produces increased salivation. Consequently, some scientists have decided that the insula is the area that processes our sense of taste.

Underlying the insula within each cerebral hemisphere is a series of masses of gray matter collectively referred to as the basal nuclei. These are located within the curve formed by the horns of the lateral ventricle. The two main structures, the lentiform nucleus and the caudate nucleus, are collectively known as the corpus striatum.

The two nuclei are separated from the insula by a sheet of gray matter covered in a layer of fiber called the claustrum. Alongside it is the biconvex shape of the lentiform nucleus (so called because it looks like a lens), which is divided into two parts. The darker and larger portion is the putamen; the smaller, lighter area is the globus pallidus (often shortened to the pallidum).

The innermost basal structure is the caudate nucleus, which forms part of the floor of the lateral ventricle. The amygdala is also sometimes included, but is more properly seen as part of the limbic system.

The basal nuclei, and their associated fiber systems, form an extremely complex set of interconnections—they are another of the brain's relay centers. Information converges on the corpus striatum from all parts of the cerebral cortex, and from the thalamus and some areas of the midbrain. The resultant output, which is almost entirely routed through the pallidum, appears to exert a strong influence on both muscle tone and motor control. This is borne out by detectable concentrations of the neurotransmitter dopamine within the basal nuclei, possibly originating in the substantia nigra of the midbrain.

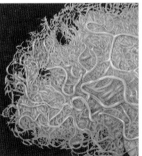

LEFT Hundreds of branching blood vessels provide a constant blood supply, essential for proper functioning of the brain.

THE HIPPOCAMPUS
The hippocampus is one part of the limbic system that is important for memory and learning.

Brain

THE OUTER LIMITS

The entire brain is sheathed by a series of three membranes known as meninges, which as well as serving to protect the delicate tissues of the brain, also provide a secure environment for the essential network of blood vessels.

The outer layer, the cerebral dura mater, is a thick, inelastic membrane which adheres to the inside of the skull, thus securing the brain. Within the dura mater run the venous sinuses, which drain blood from the brain.

The middle layer, the arachnoid, is thin and transparent, and separates the dura mater from the subarachnoid space, which contains cerebrospinal fluid and the larger blood vessels of the brain. The fluid is tapped off from the main supply at the fourth ventricle and maintains a uniform pressure on the brain, supporting it and distributing the weight evenly. The brain literally floats; a brain that weighs 3 lbs. (1.3 kg) in air, weighs about 2 oz. (56 g) when suspended in cerebrospinal fluid.

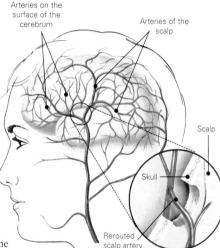

Arteries on the surface of the cerebrum

Arteries of the scalp

Scalp

Skull

Rerouted scalp artery

ABOVE Bypassing a blockage restores the vital supply of blood and oxygen to a starving brain. With a piece of plastic tubing to keep it in shape, a brain artery receives a donor artery from the scalp and the two are sewn together with microstitches. The diagram pinpoints the location of the bypass, and shows how the artery from the scalp is routed through a hole bored through the skull at precisely the correct place.

THALAMUS
The thalamus receives sensory information and relays this information to the cerebral cortex.

Brain

FOOD FOR THOUGHT

The billions of cells in the brain, from the twinkling galaxies of the cortex to the deep recesses of the basal nuclei, whether they are complex and highly specialized or just links in the chain of command, all have one thing in common. They need blood to function, and without it they die never to be replaced. Blood brings the essential oxygen and carries away the unwanted waste product, carbon dioxide. Without a constant and uninterrupted supply of blood to the brain, a human being stops moving, stops thinking, and very soon stops living.

Blood is brought to the brain by the internal carotid and vertebral arteries. The vertebral arteries join at the base of the pons, forming the basilar artery, then diverge again, forming the two posterior cerebral arteries. Each carotid artery terminates by dividing into the anterior and middle cerebral arteries. These six cerebral arteries, which supply the forebrain and midbrain, are connected by communicating arteries that form an arterial loop—the circle of Willis—at the base of the brain.

The cortex is supplied by the cortical branches of these arteries, which divide within the pia mater into smaller vessels that drop vertically into the gray matter. The underlying white matter is served by the medullary arteries; these pass through the cortex and penetrate up to 2 in. (5 cm) into the brain. The cerebellum, pons and medulla oblongata are supplied with blood mainly from the cerebellar arteries.

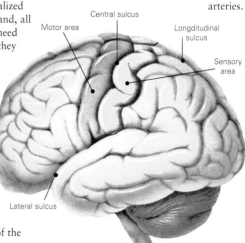

Central sulcus

Motor area

Longditudinal sulcus

Sensory area

Lateral sulcus

ABOVE This diagram separates the motor cortex from that of the sensory cortex. The two "stripes" are located on the whole brain. In sensory terms, the most sensitive body areas are the lips and the hands, with a significant region allocated to each finger. These highly mobile areas are also well supplied with coverage in the motor cortex, which also acknowledges the importance of the ankles and feet.

MOVEMENT

The basal ganglia are a group of structures, including the globus pallidus, caudate nucleus, subthalamic nucleus, putamen, and substantia nigra, that are important in coordinating movement.

Brain

BRINGING IT ALL BACK HOME

Within the brain, blood is delivered to the cells by a fine mesh of capillaries, then begins its journey back to the heart. The veins of the brain, many of which run alongside the arterial vessels, are thin walled and contain no valves. They drain into a network of venous sinuses contained within the thickness of the dura mater.

The superior sagittal sinus, which grooves the bones of the skull along its centerline, also receives blood from the nose and scalp. The inferior sagittal sinus runs along the fissure dividing the two cerebral hemispheres. Both drain down into the transverse sinuses, which continue downward to become the jugular veins. The cavernous sinuses, above the nose on each side of the head, are crisscrossed by tiny filaments and have a spongy structure. There is no lymphatic system in the brain because the tissues do not require this form of drainage.

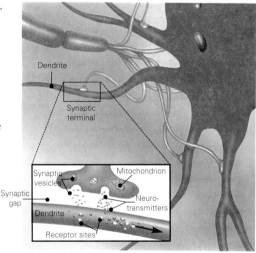

BELOW Fine tendrils from the end of a nerve axon latch on to the dendrites of neighboring nerve cells with their terminal buttons. In the synaptic cleft, at the point of contact, vesicles release neurotransmitters, which stimulate the receiving cell and pass on the nerve impulse.

Dendrite

Synaptic terminal

Synaptic vesicles

Mitochondrion

Synaptic gap

Neuro-transmitters

Dendrite

Receptor sites

THE MIDBRAIN

This area of the brain includes structures such as the superior and inferior colliculi and red nucleus. It is responsible for the function of vision, audition, eye movement, and body movement.

SMALL WONDER

Interruption of the blood supply to any part of the brain can cause serious damage. A blood clot may block an artery, or a weak-walled section may balloon out into an aneurysm, which may then burst and cause a stroke. If not immediately fatal, a stroke can cause lasting impairment and the victim may be

Brain

left without speech and minus motor functions on the side of the body opposite to the side of the brain that is affected.

Thanks to advances in microsurgery, doctors can now restore lost blood supply through bypass techniques which involve rerouting another artery. The exact location of the blockage can be determined by injecting the blood with special dyes that show up on X-rays. The neurosurgeon then selects a suitable artery in the outer scalp, close to the blockage, and which has the same diameter as the blocked vessel. Cutting through the skull with a miniature cylindrical saw, the brain artery can be exposed by cutting and peeling back the dura mater. The bloodflow in both arteries can then be stopped by applying non-crushing microclips.

Peering through a 20-power microscope, the surgeon snips through the scalp artery, and makes an oval hole in the brain artery. The supply end of the scalp artery is fed through the hole in the skull. Using a needle as fine as a baby's eyelash, it is then sewn to the brain artery with about 20 minute stitches. When the clips are removed the supply of blood is restored.

Similar techniques can be used on aneurysms, enabling the swollen arterial wall to be collapsed gently before it bursts.

WEIGHT OF BRAIN
The weight of an average brain in an adult human is between 35–36.5 oz. (1,300–1,400 g), in a newborn baby it weighs 12–14 oz. (350–400 g).

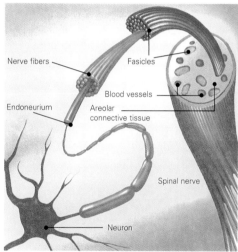

Nerve fibers

Fasicles

Blood vessels

Endoneurium

Areolar connective tissue

Spinal nerve

Neuron

ABOVE Bundles of axons form nerve fibers, which may be sheathed in a layer of myelin to form the major nerves linking body and brain.

QUARTER-INCH MIRACLE

The threat of blood starvation thus averted, the brain can continue working its normal everyday miracles. Thousands of tiny messages flashing across the cerebral cortex enable you to pick up this book, turn the pages, and read.

Brain

If the quarter-inch thick cortex were lifted off the cerebral hemispheres, unfolded and spread out, it would have an area of about two and half square feet and weigh 20 ounces or so. Like other gray matter, the cortex does not consist solely of neurons. It is in fact an intricate blending of nerve cells and fibers, neuroglia (non-excitable nerve cells that form a protective coating around nerve fibers and perform other specialist functions), and blood vessels.

MAPPING HUMAN FEELING

Although scientists can count the numbers, and describe the types, of neurons in the cortex, they cannot yet be linked in a coordinated scheme of interaction. Away from the microscope, understanding of the various surface regions of the cortex is much more precise.

The sensory cortex, spanning the front of the parietal lobe, receives information about temperature, pressure, and pain from all parts of the body. Because this information is confined to the uniform sense of touch (even from inside the body), and excludes input from specialized sense organs, it is also referred to as the "somasensory" region (from the Greek word "soma" meaning "body"). Painstaking experimentation has shown that information from particular parts of the body is received at very localized areas of the sensory cortex. Mapping

these areas has also shown that the most sensitive parts of the body (containing the most nerve endings), such as the lips and fingers, are allocated a proportionally greater area of cortex than, say, the arms or legs.

Running across the frontal lobe, parallel to the sensory cortex, is the motor cortex. Massed within its folds are batteries of motor neurons ready to fire,

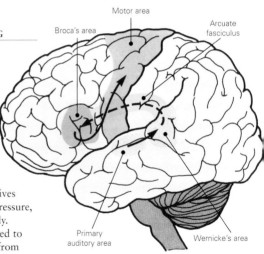

"**What is your name?**" is a common enough question, and although you do not have to think about it, providing the answer involves a sequence of almost instantaneous interconnected events in the brain. First, the question is registered in the primary auditory area and structured in the neighboring Wernicke's area. The signals then pass forward, via the arcuate fasciculus, to Broca's area, which finally passes them to the motor area, which issues instructions to the muscles of the mouth and throat to speak the answer.

Brain

initiating every movement from a handshake to a shrug. The motor cortex also exhibits the same high degree of localization and, not surprisingly, the distribution of body areas closely follows that of the sensory cortex. The area devoted to the hands and fingers reflects man's high degree of manual dexterity, and the proportion controlling the mouth and lips indicates the importance of the spoken word. Because the motor cortex must also take account of information from other parts of the brain, especially from the visual cortex at the back of the occipital lobe, it is served by a correspondingly high number of fiber tracts.

There are approximately ten billion neurons in the brain and at least a million miles of fibers. A single incoming projection fiber may serve an area of cortex containing up to 5,000 neurons, each of which can potentially interconnect with up to 4,000 others. The number of possible interconnections thus runs into trillions.

The simple task of threading a needle, which involves precise motor coordination between two hands and constant monitoring by the eyes, probably involves most parts of the brain. A detailed description of exactly what happens lies many years in the future, if ever. At present knowledge is limited to an understanding of how neurons communicate, and how a single pathway may be selected from the potential billions.

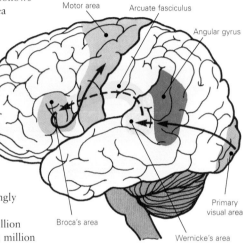

"Read this caption out loud" is a request that initiates a different sequence of events. As you scan the words, the nerve impulses triggered off the retinas in your eyes pass along the optic nerves to the primary visual area at the very back of the brain. Signals then pass for organization to the angular gyrus, before following a similar route as the auditory pathway through the arcuate fasciculus and Broca's area to the motor cortex. Once again the speech muscles are activated as you begin to speak the words.

THREE MORE FACTS
The brain can remember 50,000 different smells, is made up 80% of water, and stops growing when you reach the age of about 15 years.

HEART

The fist-sized lump of muscle, valves, and tubing that forces blood around your body is not the shape popularized on Valentine cards and representative of romantic interludes. Neither is it bright red, the color associated with such symbolism. And if it were to be run through by Cupid's arrow, the result would not be a star-struck lover—but almost certain death.

Recent times have seen the heart's place as a symbol of romance tempered by scientific reality. To the ancients, the heart contained the spiritual essence of the human being. The Egyptians weighed the hearts of the dead to measure truth. The Greeks saw the heart as a forge, burning impurities from the blood. Today, its disguise has been dispensed with, its cover has been blown. The heart is no more, no less, than a pump. But what a pump! Never resting, it thumps at least once each second (often more) from about the fourth week of conception until a couple of minutes before death. Each thump sends life-giving blood surging round the body. If each and every "lub-dub," the classic sound of a heartbeat, was written as a word in a book, you would have ten million pages in your hands. Read each "lub-dub" in one second . . . and the book would last a lifetime.

SQUEEZES AND PUMPS

Your heart contracts and relaxes some 70 or so times a minute at rest—more if you are exercising—and squeezes and pumps blood through its chambers to all parts of the body.

FATHER OF THE HEART

For 1,500 years, the heart remained shrouded in myth and mystery. Galen, the great physician of ancient Greece, assigned to it the role of a furnace, sucking in blood and combusting it to generate the heat that warms the body. Leonardo da Vinci correctly recognized the heart as made up mostly of muscle, but was also of the opinion that its job was to generate heat in some way or another.

Galen's teachings remained barely questioned until the early seventeenth century. An English physician, William Harvey, began to query the "Galenic religion" followed by doctors of the time. Through observation and experiment, on animals and human corpses, he followed a reasoned series of steps to argue that blood did not move as the ancients contended. The organ did not suck blood in, but squeezed it out. Blood did not flow back and fourth in the vessels like a tide, but away from the heart through arteries and back to it via veins. Blood could not be combusted or "used" in some other way: while inside the heart the rate of flow in and out was far too great.

Harvey came to the now obvious conclusion that the circulatory system was

Heart

by and large a closed circuit. The heart was the pump, and it drove the same blood round and round. In Frankfurt in 1628, after 12 years of experiments with more than 80 species of animals, as well as living studies and post-mortem examinations on human corpses, he published Exercitatio Anatomica de Motu Cordis et Sanguinis in Animalibus (Anatomical Treatise on the Movement of the Heart and Blood in Animals). It had an instant and profound effect.

The main gap in Harvey's proposals was that he could not explain how blood got from arteries to veins. This was the pre-microscopic era, and without a microscope, the tiny capillaries that make this connection were still invisible.

The impact of William Harvey's work on medicine cannot be overemphasized. As with the Darwinian revolution in biology, it is difficult today for us to imagine how things were ever different. Within a generation, the notion of "Harvey's heart" had been absorbed into

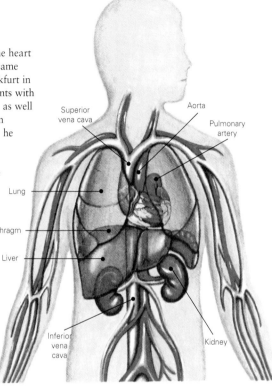

the mainstream of anatomy and physiology, and the scientific era of medicine had begun.

ABOVE Located in the middle of the chest and almost surrounded by the lungs, an adult heart is a muscular organ a little larger than a man's fist and weighing nearly a pound. It rests on the domed diaphragm where it passes over the liver. Blood passes to the heart along two main veins: the inferior vena cava, carrying blood from the lower body, and the superior vena cava, which brings in blood from the arms and head. After being oxygenated in the lungs, blood leaves the heart along the aorta, the body's main artery, which branches to serve the head and arms before arching downward to carry blood to the rest of the body.

CARDIOVASCULAR
The cardiovascular system, your body's delivery system, is made up of your heart, blood, and blood vessels.

Heart

THE HEART IN THE WOMB

The importance of the heart to survival is indicated by the speed at which it develops after conception. Every cell in the body needs a continuing supply of oxygen, and of the nutrients that provide the energy and raw materials for cellular metabolism. This is true of a tiny embryo, gradually evolving human form in the womb, as it is of a marathon runner. The oxygen and nutrients are brought by the blood, a physical, fluid conveyor belt driven around the body by heart power. In embryological development, therefore, the heart and its major blood vessels are formed and functioning long before any other organ.

The heart begins as two tiny tubes lying near each other in the primordial chest region, cross-connected by several even smaller vessels and surrounded by a sheath of muscle. Three weeks after conception, the tubes begin to fuse in the middle of the sheath. After a few days they have joined along their length to create a single, continuous chamber. By the fourth week of pregnancy this minute pocket of muscle, less than a twenty-fifth of an inch long, is folding and twisting—and has started to beat. Chambers form, marked on the heart's outer surface by grooves called sulci. The organ becomes S-shaped, and the single ventricle (lower chamber) splits into two. Until now the blood has (like the ancient teachings) ebbed

EACH DAY
Each day the average heart "beats," or expands and contracts, 100,000 times and pumps about 2,000 gallons of blood.

and flowed, but rapidly it assumes a one-way thrust around the budding circulatory system.

By the fifth week, the squirming embryonic heart has almost taken on the basic U-shape of the adult heart. The upper chamber, the atrium, divides into two. Each atrium connects with the ventricle on the same side, to create not one pump, but two. Thus the double nature of the heart comes into being. During the sixth and seventh weeks the

BELOW By the time the fetus is eight weeks old, it possesses a tiny heart that is a miniature version of an adult's. It pumps blood to and from the placenta along the umbilical cord, collecting oxygen and getting rid of wastes.

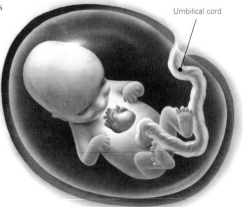

Umbilical cord

Heart

blood vessels coil and connect, and the valves that ensure a one-way blood flow are miraculously molded in living tissue.

It is the end of the second month of pregnancy. The tiny form is now nearly an inch long, and recognizably human. It is no longer termed an embryo: its correct name is a fetus. In the fetal chest, the heart pulses steadily. All four of its chambers, linked two-by-two to form adjacent left and right pumps, and separated by a tough dividing membrane known as the septum, are distinct. The valves are working. The beat is on.

Before birth, the blood circulation must be modified because the baby's respiratory and digestive systems do not work, while the placenta does. There is no breathing; the fluid-filled lungs cannot absorb oxygen from the air. The heart has a "hole" that allows communication between its left and right sides. Blood that would otherwise flow to the lungs is short-circuited back around the body, to the placenta, which is the source of oxygen (diffusing in from maternal blood) at this stage. By birth, the hole in the heart has closed. Stimulated by the newborn's first breaths, various other changes occur in the heart and circulation, the adult pattern is established and the infant is finally made physically independent of its mother by the cutting of the umbilical cord.

1
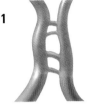

2

3

4

5

ABOVE and LEFT
An unborn baby's heart develops in stages by the fusion of the embryo's main vein and artery where they pass close to each other in the chest cavity. Three weeks after conception (1) the two tubes link together. They fuse to form a single chamber (2), before the upper atria and lower ventricles begin to take shape (3). Beginning in the fifth week (4), the atrioventricular canal is gradually split by growing ridges of tissue to provide two separate pathways for blood, one through each side of the heart. When the septum forms down the center (5), the heart has four separate chambers.

IN A LIFETIME
In an average lifetime, the human heart beats more than 2.5 billion times.

Heart

HALF A HEART EACH

Each side of the heart is a self-contained pump. Its atrium (upper chamber) is thin-walled and distensible, and receives blood flowing in from the main veins. The blood passes from the atrium through a valve to the ventricle (lower chamber), which is the muscular, thick-walled, pumping part. This contracts to squeeze blood out through another valve into a main artery.

The right pump is like the right hand—on the left, as you look at the body from the front. It drives blood to the lungs for oxygenation. This "refreshed" blood returns to the left pump, which sends it around the rest of the body to deliver the oxygen. The blood returns to the right side, completing a figure-eight.

TIRELESS MOVER

Here are some of the heart's vital statistics. It beats on average 70 times

per minute while at rest. This adds up to 100,000 heartbeats each day, and about 2.5 billion in a lifetime. But the true number is greater than this, because for a major part of each day the body is active,

BELOW The heart muscle has to have its own blood supply in order to work, and this is provided by the coronary arteries, which branch off the aorta. Cardiac veins return the "used" blood to the vena cava.

Trachea · Aortic arch · Pulmonary artery · Pulmonary veins · Superior vena cava · Left atrium · Right atrium · Left coronary artery · Great cardiac vein · Left ventricle · Anterior interventricular artery · Right coronary artery · Right ventricle · Anterior cardiac veins · Descending aorta · Inferior vena cava

THE RIGHT SIDE
The right side of the heart receives dark, bluish blood from the superior and inferior vena cava. It pumps this blood to the lungs where carbon dioxide is removed and oxygen is picked up.

Heart

and the heart has to pump faster in order to increase the supply of blood and satisfy the demands of the muscles.

An average person's body contains eight to ten pints of blood. At rest, a typical heartbeat expels about two and a half fluid ounces. This represents a theoretical cardiac output of roughly 8 to 16 pints of blood per minute, depending on the circumstances: up to 2,600 gallons each day—enough to fill a small road tanker. Again the actual volume is greater since the body does not rest all day. After strenuous exercise, the cardiac output may be ten gallons per minute.

CONE IN THE CHEST

The average adult human heart is roughly cone- or pear-shaped. It is about the size of a man's fist, measuring nearly 5 in. (12.5 cm) long, about 3 in. (7.6 cm) from side to side, and 2½ in. (6 cm) front to back. It sits in the lower chest, slightly off-center, with two-thirds of its bulk to the left of the midline. Its pointed end, or apex, is directed to the left, slightly forward and downward, so that the heart is in fact almost sideways on, with its right half facing the front.

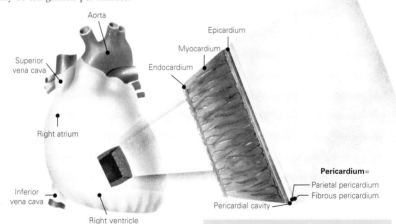

Aorta

Epicardium

Myocardium

Superior vena cava

Endocardium

Right atrium

Pericardium=
Parietal pericardium
Fibrous pericardium

Inferior vena cava

Pericardial cavity

Right ventricle

ABOVE A tough outer sheath, the pericardium, surrounds the heart and forms the outer of the three main layers that make up its walls. The middle layer is cardiac muscle, which is lined with the watertight membrane of the endocardium.

THE LEFT SIDE
The bright red oxygenated blood returns to the left side of the heart which pumps it out through the aorta to be distributed by smaller arteries to all parts of the body.

Heart

The atria, often called "upper" chambers, are in reality around the back of the heart.

A shallow groove, the coronary sulcus, encircles the outer surface of the heart and marks the division between the atria and ventricles. On the front and back, two other channels, the interventricular sulci, designate the left and right ventricles. The sulci are padded by fat and guide blood vessels around the heart itself, so that overall the organ has smooth, rounded contours.

The heart is flanked by the lungs and major blood vessels, and its apex rests on the dome-shaped sheet of muscle, the diaphragm, which forms the chest floor below. It is well protected in a "cage" formed by the breastbone (sternum), ribs, and spinal column, with their associated muscles and ligaments.

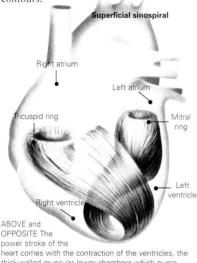

Superficial sinospiral

Right atrium

Left atrium

Tricuspid ring

Mitral ring

Left ventricle

Right ventricle

ABOVE and OPPOSITE The power stroke of the heart comes with the contraction of the ventricles, the thick-walled muscular lower chambers which pump blood either to the lungs or into the body's main circulation. The secret of their action is the way the muscle fibers spiral round the chambers, and include donut-shaped rings of muscle round the tricuspid and mitral inlet valves, to lock them tightly shut during the power stroke and prevent back flow.

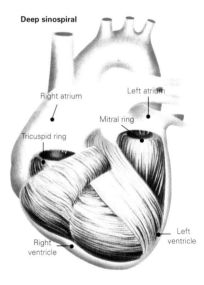

Deep sinospiral

Right atrium

Left atrium

Mitral ring

Tricuspid ring

Right ventricle

Left ventricle

BLOOD FLOW
Blood flow in the heart is controlled by valves. The tricuspid valve controls flow from right atrium to right ventricle; the mitral valve controls flow from left atrium to left ventricle.

Heart

COVERINGS AND LININGS

Wrapping the heart from base to tip is a covering known as the pericardium. Its tough outer sac, the fibrous pericardium, forms a protective sheath and is tethered by ligaments to the breastbone, spine, and other parts of the chest cavity, firmly anchoring the heart in position.

Inside the fibrous pericardium is a thin but tough double-membrane. It covers the heart's surface as its outer layer, the epicardium, and folds back on and around itself to form the parietal pericardium, which lines the fibrous portion. There are a few drops of pericardial fluid in this "double-bag," in the narrow pericardial space between the two layers of membrane. Without this the heart's powerful

pulsing movements would mean considerable friction and disruption, but the lubrication offered by the pericardial sac allows it to thump away in a kind of frictionless bath.

Inside the heart, lining its chambers and covering its valves, is the endocardium, another remarkable tissue. This glistening layer of cells must withstand the considerable internal hydraulic pressure of each heartbeat. It must also "blood-proof" the heart, so that it does not leak, by generating an ever-renewing barrier as blood rushes past within.

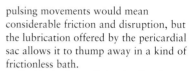

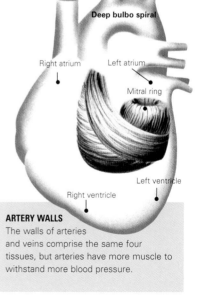

Deep bulbo spiral

Right atrium

Left atrium

Mitral ring

Right ventricle

Left ventricle

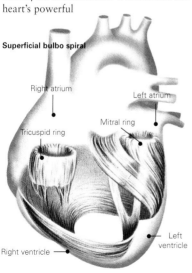

Superficial bulbo spiral

Right atrium

Left atrium

Tricuspid ring

Mitral ring

Right ventricle

Left ventricle

ARTERY WALLS
The walls of arteries and veins comprise the same four tissues, but arteries have more muscle to withstand more blood pressure.

Heart

THE MUSCLE THAT NEVER FATIGUES

Between endocardium and epicardium is the heart's powerhouse: a layer of heart or cardiac muscle, technically termed myocardium. Spiraling bands of muscles wrap themselves around a framework of dense, fibrous tissue that forms the "skeleton" of the heart. (The fiber skeleton and the valves inside the heart make up about half of the organ's weight.) The muscle bands curve around each chamber to form its wall, and they also coalesce into loops which support the valves that separate the atria from the ventricles.

The atrial walls have much less cardiac muscle than the ventricles and so are quite thin, the left atrial wall being the more substantial. These chambers are not so much power pumps as expandable reception chambers for incoming blood (atrium is Latin for entrance hall). The left ventricular wall, which produces the pressure to drive blood all the way to the fingers and toes and back again, has the greatest muscular mass—it measures up to half an inch thick in places. The right ventricle's task of circulating blood through the lungs is less arduous; this chamber sits on the side of the left ventricle, and its walls are less than a quarter of an inch thick.

Cardiac muscle is a curious hybrid of the other two main muscle types in the body, skeletal and visceral muscle. Under the microscope, cardiac muscle is seen to have the stripes, or striations, of skeletal muscle, the type that moves the bones of the skeleton and is largely under voluntary or conscious control. Yet heart muscle is not under conscious control: it responds to the part of the nervous system called the autonomic or involuntary nervous system, as well as to its own internally generated electrical commands. In this respect cardiac muscle is more like the visceral, or smooth (stripe-less), muscle which lines the stomach and other internal organs.

The fibers of cardiac muscle are in effect enormous cells, tiny fractions of an inch long. Like skeletal and visceral muscle, they contain bundles of actin and myosin filaments, and in respect of the molecular basis of contraction they resemble other muscle tissue in the body. They also have an abundance of mitochondria, the cellular power centers that convert food into energy. But they differ from other muscles in the way that electrical signals, constituting a nerve message, travel through the mass of fibers.

FROM AN ELEPHANT TO A MOUSE
While an average person's heart beats between 60 and 80 times a minute, by comparison, an elephant's heart only beats 20 to 25 times per minute and the heart of a mouse beats more than 500 times each minute.

Heart

Where two cardiac fibers meet end to end, there are distinct dark bands known as intercalated disks. At certain points along each disk, the outer membranes of the two adjacent cells fuse, so that the two cardiac fibers share the same membrane at these so-called "tight junctions." Electrical signals flow almost unimpeded through the tight junctions. But at other points along the cell membrane the electrical resistance is hundreds of times greater. So the signals follow the line of least resistance and hop from one fiber to the next, leaving in their wake a chain of contractions. Structurally, cardiac muscle is a lattice work of separate cells. Functionally, it behaves as a syncitium—a group of cells that have merged to act with common purpose, like a single giant cell.

IN ONE DAY

In one day, your heart pumps enough blood to fill the size of a truck that delivers heating oil to homes.

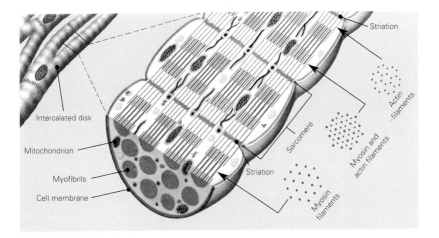

Striation

Actin filaments

Myosin and actin filaments

Myosin filaments

Sarcomere

Striation

Intercalated disk

Mitochondrion

Myofibrils

Cell membrane

ABOVE Dissected lengthwise, a cardiac muscle fiber reveals the origin of its banded structure. Parallel bundles of myofibrils are made up of sections called sarcomeres, which consist of overlapping filaments of the proteins actin and myosin. The filaments slide over each other to make the fiber contract, and the combined effect of thousands of such contractions maintains the pumping action of the heart.

Heart

OPEN AND SHUT

Working within the heart's tireless muscles are four equally durable valves. They open and shut in a set order every time the heart contracts, ensuring blood flows in the correct direction around the system, rather than the same portion simply being sucked in and ejected with each beat. The four valves are in two pairs; atrioventricular valves between atria and ventricles, and semilunar valves between ventricles and main arteries. Each is sculpted from tough, rubbery flaps of fibrous tissue, the flap being sandwiched between two layers of endocardium and anchored to the tough rings of the muscle and fibrous skeleton.

The tricuspid valve guards the entrance to the right ventricle, ensuring that blood comes in from the right atrium but does not exit that

way. As its name suggests, it has three tooth-shaped flaps, or cusps. The mitral valve does the same job for the left ventricle; it possesses two unequal cusps. It is named, rather less scientifically, from its resemblance to a bishop's miter.

Each of these atrioventricular valves works in a supremely simple way. As the ventricle contracts, the blood inside is pressurized and it pushes up on the undersides of the cusps, forcing their edges together and pinching them shut to create a seal. Long, thin tendons (chordae tendineae) lead from the edges of the cusps down through the ventricular chamber and are anchored to muscles in its sides; these prevent the cusps from being turned "inside out" and forced through into the atrium. As the ventricular muscle relaxes and the ventricle expands, the cusps swing back against its wall and so allow blood to surge through from the atrium.

While the atrioventricular valves are closing, the semilunars are opening. These valves were christened from their three crescent-shaped cusps, which are hollow and pouchlike. The pulmonary valve permits blood to flow from the right ventricle out into the main artery, the pulmonary artery, on its route to the lungs. The aortic valve fulfills the same function for the left ventricle, letting blood surge out into the body's main artery, which is called the aorta.

These two valves also operate in a simple way. Blood flowing the correct way pushes the cusps flat against the wall.

THE AORTA
The aorta is the largest artery in the body, and is almost the diameter of a garden hose. Capillaries, on the other hand, are so small that it takes ten of them to equal the thickness of a human hair.

Heart

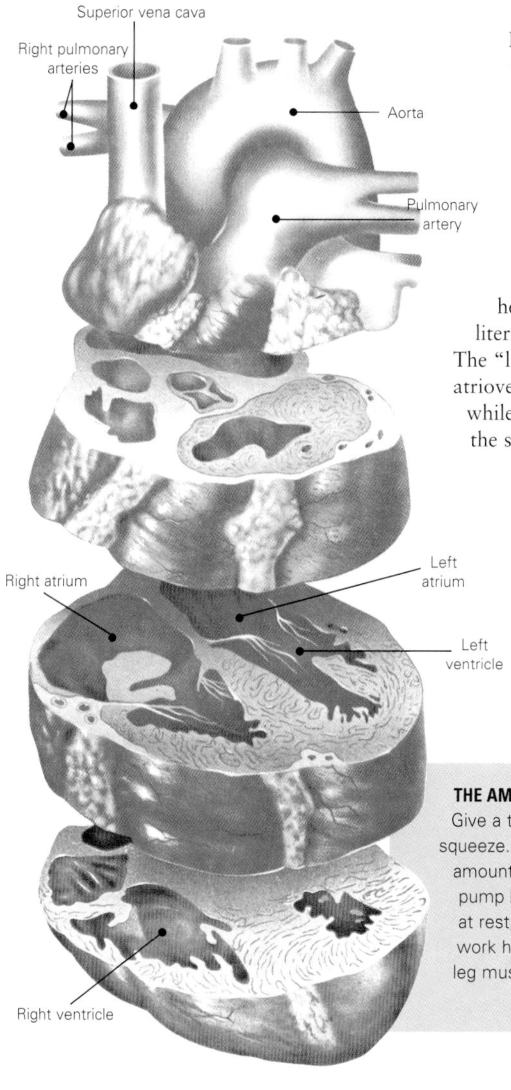

Superior vena cava

Right pulmonary arteries

Aorta

Pulmonary artery

Right atrium

Left atrium

Left ventricle

Right ventricle

Blood attempting to flow the wrong way fills the cusps and balloons them out so that their edges come together to make a seal.

As the valves slap shut to prevent blood's backflow, they make a noise. You have probably heard it: the "lub-dub" of a heartbeat, immortalized in literature, theater, and romance. The "lub" represents the atrioventricular valves closing, while the "dub" is the sound of the semilunar valves shutting.

LEFT The heart's chambers vary in size and their walls vary in thickness, as shown by the cross-sections. Most noticeable is the thickness of the muscle that makes up the ventricle walls. The left ventricle (which contracts to pump arterial blood out of the heart) is more powerful than the right.

THE AMOUNT OF FORCE
Give a tennis ball a good, hard squeeze. You're using about the same amount of force your heart uses to pump blood out to the body. Even at rest, the muscles of the heart work hard—twice as hard as the leg muscles of a person sprinting.

Heart

THE CROOKED CROWN

All tissues of the body need a blood supply, especially active muscle. The heart is active muscle, and each of its fibers is paralleled by a capillary bringing bloodborne oxygen and nutrients. Paradoxically, the organ cannot use the blood coursing through its chambers. The rate of flow is too fast, and the internal pressure too great; they would rupture the delicate network of cardiac capillaries. In any case, blood in the right side of the heart is poverty-stricken as far as oxygen is concerned—it would be unable to supply this vital substance to the fibers there.

So the hollow muscle that is the heart keeps 5 percent of the blood it pumps (only the brain takes a greater supply). This organ, filled with blood, maintains its own supply from the outside. Two coronary arteries branch from the main aorta just above the aortic valve. No larger than drinking straws, they divide and encircle the heart to cover its surface with a lacy network that reminded physicians of a slightly crooked crown (coronary comes from the Latin coronarius, belonging to a crown or wreath). They carry about

130 gallons of blood through the heart muscle daily.

The left coronary artery divides into the anterior descending coronary artery, which carries blood down the front of the heart to both ventricles, and the circumflex artery, which winds around the back to nourish the left ventricle and atrium. The right coronary artery curves round and down to send one branch, the marginal artery, to the fronts of the right atrium and ventricle; a second branch, the posterior descending coronary artery, nourishes the backs of both ventricles.

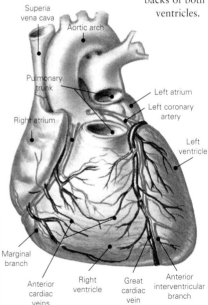

Superia vena cava
Aortic arch
Pulmonary trunk
Left atrium
Left coronary artery
Right atrium
Left ventricle
Marginal branch
Anterior cardiac veins
Right ventricle
Great cardiac vein
Anterior interventricular branch

LUB-DUB
Lub-DUB, lub-DUB, lub-DUB. Sound familiar?...If you listen to your heartbeat, you'll hear two sounds. These "lub" and "DUB" sounds are made by the heart valves as they open and close.

Heart

The coronary arteries are mirrored by a system of cardiac veins which carry blood back from the muscles of the heart. These veins generally parallel the main arteries, to join at a small chamber, the coronary sinus, which empties into the right atrium.

TWICE AROUND THE BLOCK

We can now trace blood's double circuit in more detail. Dark venous blood, low in oxygen and high in waste carbon dioxide, is squeezed out of the right ventricle, through the pulmonary valve and into the pulmonary artery. Shortly this branches right and left, one branch supplying each lung. In the lungs, the arteries continue to branch repeatedly until they become capillaries, microscopic blood vessels which encircle the millions of tiny air sacs (alveoli).

Here the blood rids itself of carbon dioxide into the alveolar air, and recharges itself with oxygen from the same, turning bright reddish-scarlet in the process.

The oxygenated blood drains along the four pulmonary veins to the heart's left atrium, through the mitral valve. It then passes into the thickest-walled, most muscular chamber of the four, the left ventricle.

As this contracts, the oxygenated blood is forcefully expelled through the aortic valve in the aorta, the chief artery with branches to all parts of the body. In the vast capillary network that permeates all tissues, blood finally donates its oxygen and takes on board carbon dioxide, again turning dark reddish-blue. The capillaries unite, eventually to form the two main veins, the superior and inferior venae cavae, that guide the blood back to the right atrium, thence through the mitral valve to the right ventricle. The circuit is complete; it takes only half a minute.

LEFT and FAR LEFT Encircling the heart like a crown, the coronary circulation brings oxygen-carrying blood to the heart muscle itself. During times of great exertion, the heart beats faster. The heart muscle's oxygen demand rises and the coronary arteries have to deliver more blood to the very organ that pumps it there.

AMAZING POWER
Your heart exerts enough power every day to lift a 1 ton weight to 41 ft. high.

Superia vena cava / Aortic arch / Left pulmonary vein / Left atrium / Coronary sinus / Circumflex branch / Middle cardiac vein / Left ventricle / Right atrium / Posterior interventricular branch / Right ventricle

Heart

THE CARDIAC CYCLE

Heartbeat is a simple cadence of contraction and relaxation. A complete beat is termed the cardiac cycle. Its contractile phase, when blood squirts out of the heart into the arteries, is systole. Its relaxation phase, in which the heart momentarily rests and draws in another gulp of blood, is diastole.

The exact nature of the contraction—how it progresses through the various chambers—is elegantly ingenious. Seen from without, as the layers of cardiac muscle contract in systole, they seem to wring the heart of blood. The contractile wave begins near the top of the right atrium and crosses both upper chambers. Then another wave of contraction arises

EXERCISE
Exercise, such as walking for four hours a week, can protect your heart. Sex is also good for your heart.

BELOW and BEOW RIGHT The "lub-dub" sound of a beating heart is caused by alternate diastolic and systolic phases (below, left, and right). The finely tuned rhythm of the cardiac cycle can be traced in this sequence of diagrams, in which non-oxygenated blood is shown in blue and oxygenated blood is shown in red. As the heart relaxes in diastole (1), both upper chambers (atria) fill with blood—non-oxygenated blood arrives in the right side from the body's main veins, and oxygenated blood returns to the left side after its trip to the lungs. The mitral and tricuspid valves open, and at systolic contraction (2) the atria force blood into the heart's lower chambers (ventricles). The ventricles then contract in their turn (3), pumping non-oxygenated blood through the pulmonary valve and on its way to the lungs, and forcing oxygenated blood through the aortic valve into the body's main circulation. The atria relax again as they re-enter diastole (4), and fill with blood once again to restart the cycle.

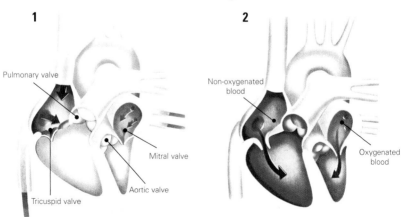

1

Pulmonary valve

Mitral valve

Aortic valve

Tricuspid valve

2

Non-oxygenated blood

Oxygenated blood

Heart

from the apex, at the bottom of the heart, and travels upward to envelop the ventricles.

A moment's consideration shows this must be so. The atria begin their contractions from the top because they funnel blood through to the ventricles; the latter squeeze from below, because blood must be ejected upward into the aorta and pulmonary artery. In systole the heart simultaneously shortens, flattens, and twists about one-quarter of a turn, thrusting its left half toward the front of the chest.

During a typical heartbeat, with a duration of around four-fifths of a second, the atria contract for about 10 percent of the time and the ventricles for about 40 percent. Ventricular contraction

YOUR PULSE
Feel your pulse by placing two fingers at pulse points on your neck and wrists. The pulse you feel is blood stopping and starting as it moves through your arteries. As a kid, your resting pulse might range from 90 to120 beats per minute. As an adult, the rate slows to an average of 72 beats per minute.

begins with a 0.06-second phase, followed by the main phase of 0.11 seconds in which 60 percent of blood output is expelled. The third phase, lasting around the same time as the second, ejects the remainder of blood for that cycle.

3

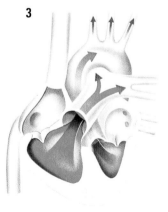

4

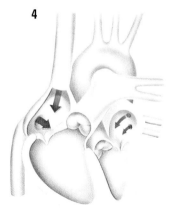

Heart

THE WIRING WITHIN

The smoothly choreographed motion of a beating heart gives the impression of order within. Electricity is the key. Millions of tiny electrical nerve signals flash around the body every second, instructing and coordinating muscular activity. The heart too has its electrical signaling system. However, it is self-contained and based, not on nerves in the strict sense, but on muscle cells modified to transmit electrical signals. Clusters of these unique cells lie buried in the heart, and the results of their teamwork are manifested as the power to pump.

The heart's tempo is set by a bundle of specialized cells high in the wall of the right atrium. Equipped not only to transmit but also to generate, these cells make up the sinoatrial, or SA node. This is the heart's own natural pacemaker. The cells create their own impulses; they "self-excite," and their excitement is infectious.

At the start of a beat, signals from the SA node flash around the atrial walls to stimulate cardiac muscle contraction. They also pass through to the septum, the muscular wall separating the two sides of the heart. Deep in the septum, close to where the heart's four chambers converge,

lies a second group of cells, the atrioventricular, or AV node.

The AV node is a relay station. It receives the signals from the SA node, which have been traveling along a highly conductive pathway in the atrial wall, at a speed of around 24 in. (60 cm) per second. In the node itself, the speed of conduction falls to only 2 in. (5 cm) per second. Then the AV node fires the signals onward along another electrical conduit, a tract of yet more specialized cardiac fibers named the bundle of His (for the German physiologist who discovered it). The signals speed up as the bundle of His forks into two branches, each spreading into a profusion of tendrils called Purkinje fibers—the "distribution nerves" of the heart. The fibers network their way through the ventricular walls, spreading the electrical messages at the rate of almost 6 ft. (3.7 m) per second, to the waiting cardiac muscle fibers.

FEEDING THE HUNGRY MILLIONS

The heart, marvel of stamina and power, is at the functional—and roughly anatomical—center of the circulatory, or cardiovascular, system. Pumping blood is all very well, but the blood needs to be delivered to where it is needed, as demands on various parts of the body change. The arteries and veins are the highways of the delivery network, while the capillaries are the byways and trading posts, feeding the millions of cells in the

CARDIOVASCULAR DISEASE
At least 58,800,000 million Americans suffer from some kind of heart disease, that is about one person in four.

Heart

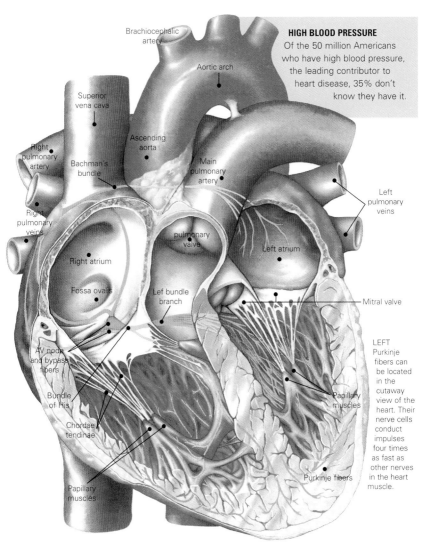

Brachiocephalic artery

Aortic arch

HIGH BLOOD PRESSURE
Of the 50 million Americans who have high blood pressure, the leading contributor to heart disease, 35% don't know they have it.

Superior vena cava

Ascending aorta

Right pulmonary artery

Bachman's bundle

Main pulmonary artery

Right pulmonary veins

Left pulmonary veins

pulmonary valve

Right atrium

Left atrium

Fossa ovalis

Lef bundle branch

Mitral valve

AV node and bypass fibers

Bundle of His

Chordae tendinae

Papillary muscles

Papillary muscles

Purkinje fibers

LEFT Purkinje fibers can be located in the cutaway view of the heart. Their nerve cells conduct impulses four times as fast as other nerves in the heart muscle.

Heart

body's every nook and cranny.

Arteries divide repeatedly to form capillaries, which unite repeatedly to form veins. The entire length of the system has been estimated at a staggering 80,000 miles (128,746 km).

The functions of the circulation are many. Oxygen, of course, and nutrients are a regular delivery order for the tissues. Carbon dioxide and other molecules of metabolic refuse are just as regularly collected up for disposal. The circulatory system also distributes heat from "hot," metabolically active organs such as the liver to cooler, more quiescent areas like resting muscles. Integrated into this aspect is a thermostatic function: divert more blood to the skin and more heat is lost to the external environment, while reserving the bulk of blood flow for the internal organs instead helps to conserve heat.

The cardiovascular system also serves a longer-term regulatory function. A country's road transport network evolves with demand: areas of increased activity receive bigger and better roads, while lack of action allows routes to fall into disrepair and decay. So with the circulation. In response to lengthy periods of elevated oxygen demand from any organ, the

blood vessels to and within that organ increase in size and number. Conversely, depressed oxygen demand shrinks the vessels in number and diameter.

FROM SURGE TO OOZE

Arteries are the vessels that guide blood away from the heart. You might think that all arteries contain bright red blood, under great pressure, which would spurt out if the artery were severed. In most cases, especially of arteries likely to be damaged in an accident, this is true. But the first assumption is incorrect. The pulmonary artery and its branches leading to the lungs carry dark reddish-blue blood, low in oxygen, and not the highly-

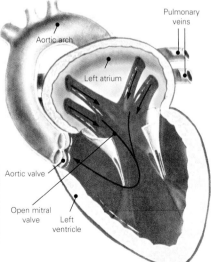

Pulmonary veins

Aortic arch

Left atrium

Aortic valve

Open mitral valve   Left ventricle

WHAT IS A HEART ATTACK?
A heart attack occurs when we develop a blockage in one of the arteries supplying blood to our heart.

Heart

oxygenated, bright red version. By the same token, the pulmonary veins transport oxygen-rich, bright scarlet blood and not the dark venous variety.

Arteries and veins have the same layers of tissues in their walls, but the proportions of these layers differ. Lining the bore of each is a thin endothelium, and covering each is a sheath of connective tissue. But an artery has thick intermediate layers of elastic and muscular fibers, whereas in a vein these are less developed. The thick, elastic arterial wall helps it withstand and absorb the pressure wave newly generated in the ventricles and hydraulically transmitted by the fluid medium of the blood. The wall expands with the swelling force of systole, then snaps back to urge the fluid on its way as the heart takes a rest during diastole; the semilunar valves prevent any backflow. In this way the pressure peaks are gradually flattened along the branches of the arterial tree.

As blood enters the capillary network, the pressure falls off. By the time it reaches the veins, the red fluid is not so much surging, as oozing.

There is no need for strength or containment here, so the venous walls are thin to the point of floppiness. To compensate for this lack of dynamic force, many veins lie sheathed in skeletal muscle. The slightest shift of a limb squeezes the vein and drives the blood toward the heart. To ensure the right direction of flow, valves are again the order of the day. Veins, particularly those in the arms and legs, have many semilunar valves. These swing open for each faint pulse of blood, then flap shut to prevent flow reversal.

WHAT IS A STROKE?

A stroke is a result of a blockage in one of the arteries to our brain. In both a heart attack and stroke, lack of blood stops the heart or brain from working so it shuts down and we collapse.

BELOW and OPPOSITE Like other muscles in the body, the heart thrives on exercise. Regular physical demands improve the strength of the heart muscle, increasing by more than twice the amount of blood pumped out with each beat—the so-called stroke volume—which can be an important asset when the body's muscles need an increased supply of oxygen.

Aortic arch

Pulmonary veins

Left atrium

Closed mitral valve

Aortic valve

Two-fifths of blood volume stays in ventricle

Left ventricle

KIDNEYS

The body has four routes for elimination of wastes. One is the skin, for removing certain salts and minerals. A second is the lungs, for getting rid of carbon dioxide and of water in vapor form. A third is the intestines, for removal of undigested leftovers from food—although in a sense, undigested remains have never been truly "inside" the body. The fourth involves the principal excretory organs, the kidneys.

THE BODY'S BALANCING ACT

The paired kidneys in the upper part of the abdomen toward the back, perform a masterly balancing act. They balance the fluid levels in the body. They balance the body's acid/alkaline nature, and they balance concentrations of salts, minerals, and other substances. In times of plenty, unwanted materials are removed; when times are hard, the kidneys conserve.

The kidneys work through the medium of the blood. Blood flows everywhere in the body, fetching and carrying, distributing and collecting. Part of its circulation is through the kidneys, and there it is balanced—filtered, purified, cleaned, and adjusted. Like a commercial effluent treatment plant it is a continuous process, 24 hours each day.

The kidneys form part of a major body system, the excretory system.

URINARY TRACT
The kidneys make up part of the urinary tract along with the ureter, bladder, and urethra. These organs work together to produce, transport, store, and excrete urine.

Kidney

Ureter

Bladder

Urethra

ABOVE The urinary system filters the blood and stores waste products in urine. It consists of the kidneys, ureters, bladder, and urethra. In addition to removing dissolved wastes from the blood, fed in along the renal arteries, the kidneys maintain the balance of salts, water, and trace elements in the body.

Kidneys

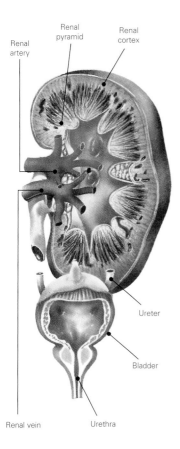

Renal pyramid

Renal cortex

Renal artery

Ureter

Bladder

Renal vein

Urethra

ABOVE Purified blood is returned to the circulation along the renal veins. Urine drains from the renal pelvis into the ureter, to be stored in the bladder. In men the beginning of the urethra, the outlet from the bladder, is surrounded by the globular prostate gland.

Its other structures include the two ureters, tubes which lead from the kidneys to the bladder, and the urethra, which leads from the bladder to the exterior of the body. Waste disposal by the system can be described relatively simply. The kidneys filter blood and produce a filtrate, urine, which contains unwanted substances. The urine is stored in the bladder until such time as it is convenient for it to be expelled during the act of urination (technically called micturition).

The simple description belies a marvel of metabolic juggling. Hundreds of pints of fluid and dozens of chemicals are involved. The renal arteries supply blood to the kidneys. The flow they carry is an astounding 750 to 1,000 pints (426 to 568 l) daily, approaching one-quarter of the heart's output, and roughly equivalent to the body's entire blood volume circulating through the kidneys 20 times each hour. However, only one-to-two thousandths of the blood flow emerges at the other end, as urine. The average daily production of urine is just over 2½ pints (1.4 l), although it varies considerably according to how much we eat and drink, how active we are, the ambient temperature, and other factors.

FUNCTION
Being both about the size of a child's fist, the kidneys most important function is to filter blood and produce urine.

Kidneys

MICROSCOPIC SIEVES

The Roman physician Galen, of the second century AD, suggested that the interior of the kidney was a sieve which filtered out impurities into the urine. He imagined it as one large sieving surface with innumerable pores too small to see. In fact, his notion was not too wide of the mark, considering that the role of the heart and the circulation of the blood were not realized until 15 centuries later, by William Harvey.

However, the gross anatomy of the kidney reveals almost nothing of its finer workings. Slice one in half and you first encounter its tough external coat, the renal capsule. Inside the organ, the smooth outer layer is known as the cortex; the more stripy, textured inner layer is the medulla, and it is grouped into lobes termed pyramids. On its concave side there is a converging system of collecting spaces, the renal pelvis, which gradually narrows to form the thin tube of the ureter.

The hidden secrets of kidney function had to wait until the invention of the microscope. Celebrated Italian microscopist Marcello Malpighi studied its tissue and published his observations in 1659. But with no wider framework of knowledge, and with the mighty Galen's "sieve" still deeply entrenched in the medical beliefs of the time, Malpighi failed to make progress. He was denounced by his colleagues for imagining what he saw through the eyepiece.

In 1842, the English physician William Bowman wrote the treatise that proved a watershed in renal studies. "On the Structure and Use of the Malpighian Bodies of the Kidney With Observations on the Circulation through that Gland." Bowman had unraveled the secret of the first part of kidney function, describing how the maze of tiny tubes and blood vessels in the organ managed to filter wastes and water from the blood. Only two years later Carl Ludwig, in Vienna, showed that this initial filtrate was then concentrated in a secondary part of the tube network, as the body reabsorbed most of the water from it. We now know that the kidneys filter up to 350 pints (199 l) of water and solutes from the blood each day, in the first part of the process; and that reabsorption of water from this filtrate brings down its volume to 2½ pints (1.4 l) of actual urine per day.

KIDNEY BEANS

Humans are symmetrical organisms. We have paired arms, paired lungs, and paired kidneys. The two kidneys are tucked in the upper abdomen, the left one

IMPORTANCE OF URINE
Urine contains the by-products of the body's metabolism: salts, toxins, and water. When you are asked to give a urine sample during a doctor's visit, the results reveal how well your kidneys are working.

Kidneys

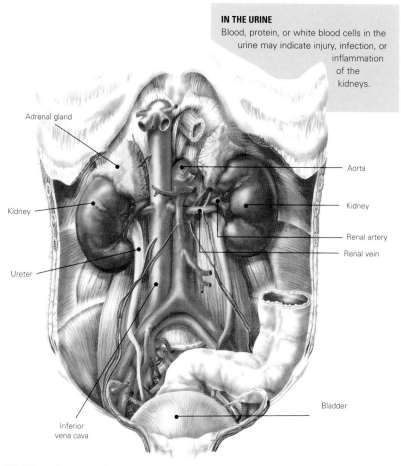

IN THE URINE
Blood, protein, or white blood cells in the urine may indicate injury, infection, or inflammation of the kidneys.

Adrenal gland

Aorta

Kidney

Kidney

Renal artery

Renal vein

Ureter

Inferior vena cava

Bladder

ABOVE For maximum protection against injury, the kidneys are tucked up toward the rear of the abdomen, guarded by the lower ribs. Their never-failing blood supply flows along the renal arteries, which branch off the aorta, the body's main artery from the heart. Blood cleansed of wastes by the kidney's filters returns along the renal veins to the inferior vena cava, on the way to the heart to be pumped back into the circulation.

Kidneys

being about ½ in. (1.2 cm) higher than the right. Many people perceive their kidneys as being lower and more to the side than they actually are. A common misconception is that they are just under the skin of the "flanks," on the sides of the abdomen, broadly at waist level. In fact they are buried much more deeply, centrally and higher, being only 2 in. (5 cm) or so from the body's midline and directly behind the lower ribs. The top of each kidney is more rounded than the base, and tilts inward toward the midline. As you stand up from a lying position, your abdominal organs "sag" and the kidneys move down with them, about an inch lower than their recumbent pose. They also bob up and down during breathing.

Behind the kidneys is the musculature of the spine; the left kidney is flanked by the pancreas, jejunum (part of the small intestine), and the colon; the right kidney, displaced downward by the bulk of the liver, touches the duodenum and colon. This array of ribs, spine, and other adjacent parts provide superior protection for such important organs. In addition, the kidneys are padded by a special type

KIDNEY DISEASE
The two most common causes of kidney disease are diabetes and high blood pressure. If your family has a history of any kind of problems, you may be at risk for kidney disease.

of surrounding fat which holds and cushions them firmly.

Atop each kidney sits an adrenal, or suprarenal, gland. These glands are not involved in excretion, being part of the body's endocrine or hormonal system.

This arrangement is the normal one. Some people have only one kidney, either through a developmental abnormality or following surgical removal of the second kidney because of disease. Or the two kidneys may develop fused together into a horseshoe shape. These departures should pose no problems, provided the existing kidney is not diseased in any way.

KEEPING THE INSIDE CONSTANT

The kidneys also cope with the variety of inputs and activities to which the body is subjected each day. You might overload your system with food one day but hardly eat the next. You might then have a very active day, moving energetically and sweating profusely without taking in fluid to compensate. The following day might be spent in restful repose, yet you might take in large amounts of fluid. Through the medium of the blood and other internal fluids, the kidneys iron out these fluctuations. They are the great levelers.

Body cells, left to their own devices, cannot tolerate much change. They are specialized and delicate devices, able to perform their singular and specific functions as part of the body's team, but unable to look after themselves in more basic ways—such as controlling their

Kidneys

BELOW The main structures within the kidney are the outer cortex, and the inner medulla. Bunches of medulla form pyramids, from which urine drains into the pelvis of the kidney, then enters the ureter.

WITH JUST ONE
Although the kidneys work in tandem to perform many vital functions, people can live a normal, healthy life with just one kidney. Some people are born with just one kidney.

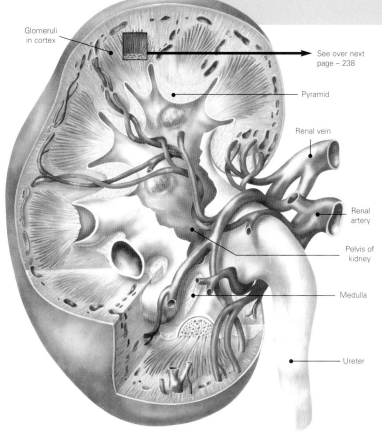

Glomeruli in cortex

See over next page – 238

Pyramid

Renal vein

Renal artery

Pelvis of kidney

Medulla

Ureter

Kidneys

temperature and immediate chemical environment. They must be cared for and cosseted. The nineteenth-century French physiologist, Claude Bernard, first recognized the presence of the milieu interieur, or "internal environment."

The body is not a series of watertight compartments, each holding an organ. Cells contain their own, intracellular, fluid. They are bathed in extracellular fluid, which undergoes a slow interchange with the intracellular fluid and with both blood and the lymph fluid of the lymphatic system. Together these fluids must provide constancy of temperature, of sugar and salt concentration, and of acidity and alkalinity. Then the specialist workers, the cells, can get on with their jobs. The kidneys are the carers, keeping the inside constant.

BELOW Millions of nephrons and their attendant blood vessels loop and twist in the cortex of the kidney. In each nephron a knot of arteries, the glomerulus, acts as a filter unit at the wide end of a loop of renal tubule that collects the fluid filtrate and dips right down into the kidney's medulla.

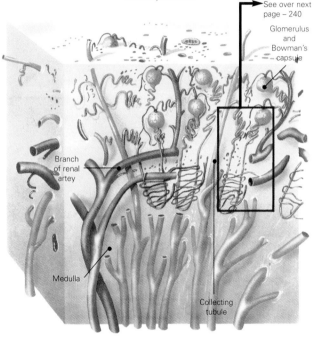

See over next page – 240

Glomerulus and Bowman's capsule

Branch of renal artey

Medulla

Collecting tubule

THE WASTE
The wastes in your blood come from the normal breakdown of active tissues and from the food you eat. After your body has taken what it needs from the food, waste is sent to the blood.

Kidneys

A MILLION MICROFILTERS

The kidney's sieves, or filters, can be seen
only with the help of a high-power
microscope. Their tiny tubes coil and
intertwine, so that early researchers had
much trouble trying to untangle the organ's
microanatomy, which has been described as
"raspberries and spaghetti." However,
today we have a clear description of the
nephron, the active site of filtration.

There are about one million nephrons
in each human kidney. Each nephron
consists of several parts. There is a tiny
knot of capillaries, the glomerulus
("raspberry"). The glomerulus is cupped
in the expanded end of a long tube. The
expanded part is known as Bowman's
capsule (in honor of William Bowman),
and the long tube (the "spaghetti") is
termed the renal tubule. It twists and
turns the vicinity of the capsule and then
throws a long loop, the loop of Henle,
from which it returns to the
neighborhood of the capsule. Finally the
tubule winds away from the glomerulus
and capsule, and joins with other tubules
to form a larger collecting duct.

The glomeruli, Bowman's capsules,
and their adjacent parts of the renal
tubule are embedded in the kidney's outer
cortex—a million sets of each. The long
loops of Henle (many are nearly 1 in. /2.5
cm long) and the collecting ducts are
found in the medulla. The ducts come
together and open into the renal pelvis,
the kidney's central space. If the tubules
of all nephrons were uncoiled and joined

end to end, they would stretch for 50
miles (80 km).

A single nephron's structure is perhaps
best detailed by following the fate of
blood as it flows along the renal artery,
heading for the kidney. In this way it is
possible to trace, step by step, the
conversion of certain elements of blood
into urine.

THE VITAL BLOOD SUPPLY

The renal artery leaves the aorta and
divides into five branches, which fan out
on reaching the renal pelvis in order to
supply five distinct areas of the kidney.
About one person in three has one or
more of these branches coming directly
off the aorta and entering the segments
within the kidney. For them, a single renal
artery as such does not exist.

The kidneys, like the heart and brain,
receive a "privileged" blood supply in
preference to other parts of the body.
Even when total blood flow varies widely,
because of dehydration or blood loss, the
flow rate through the kidneys remains at
roughly 120 pints (68 l) per hour. This
proves their importance to the body: their
job is so vital that it must continue, even
if this means putting other systems under
stress.

Kidneys

Once inside the kidney, the arteries branch again and again as they travel up the pyramids of the central medulla, until they reach the edge of the cortex. There they fan out, and smaller branches lead off at right angles to feed every part of the cortex. These tiny interlobular arteries give rise to the even smaller arterioles, which go on to form the tangles of capillaries, the glomeruli. The initial filtration of the blood is about to begin.

THE FILTERING SURFACE

The actual filtering surface is the glomerular basement membrane, which acts as a selective barrier between the capillaries of the glomerulus itself and the beginning of the urinary space within the Bowman's capsule. The basement membrane is made of a mat of filaments which, on the capsule side, interlock with each other

THE TUBULES
The actual filtering occurs in tiny units inside your kidneys called nephrons. In the nephron, a glomerulus intertwines with a tiny urine-collecting tube called a tubule.

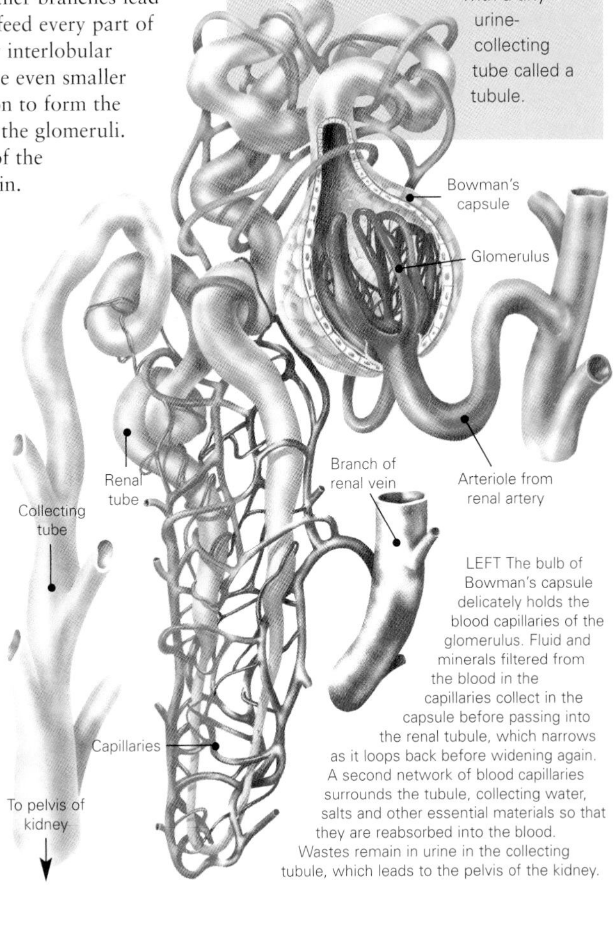

Bowman's capsule

Glomerulus

Renal tube

Collecting tube

Branch of renal vein

Arteriole from renal artery

Capillaries

To pelvis of kidney

LEFT The bulb of Bowman's capsule delicately holds the blood capillaries of the glomerulus. Fluid and minerals filtered from the blood in the capillaries collect in the capsule before passing into the renal tubule, which narrows as it loops back before widening again. A second network of blood capillaries surrounds the tubule, collecting water, salts and other essential materials so that they are reabsorbed into the blood. Wastes remain in urine in the collecting tubule, which leads to the pelvis of the kidney.

Kidneys

like the bristles of two hairbrushes pressed together. Gaps or "slit pores" between the bristles allow comparatively small molecules to pass through. Nothing as big as a red blood cell can normally traverse the membrane from blood to the urinary space.

The fluid pressure of blood in the capillaries is much higher than that of fluid in the urinary space (the journey from the heart, where the pressure originates, to the kidney is very short and blood still retains much of its force). Plasma, the watery part of the blood which contains many dissolved substances, is pushed through the glomerular basement membrane by this irresistible pressure gradient. With it travel smaller molecules such as urea (the nitrogenous waste product), creatine (a waste from muscle metabolism), glucose, mineral salts, and some of the small blood proteins. It is, in effect, a pressure-filtration system. Bigger molecules and blood cells stay in the capillary.

The basement membranes offer a large surface area for filtration. In a normal day, for an average adult, about 350 pints (199 l) of fluid from food and drink are filtered from the glomeruli of both kidneys into the tubules. This means that 2½ fl. oz. (0.071 l) of liquid pass through the collective basement membranes of one kidney each minute.

NEPHRONS
Every kidney has about a million nephrons.

SUPER SELECTIVITY

The liquid that comprises this initial filtrate contains many substances that the body cannot afford to lose, including glucose (its chief energy source), amino acids (the building-blocks of proteins), and mineral salts such as sodium, chlorides, and phosphates. Water itself also passes through the filters in abundance, and the body cannot risk losing it at such a rate. So, after the initial "blunderbuss" of filtration, comes selective reabsorption. It is rather like trying to sort items in a large box. The renal method would be first to tip them all out on to the floor (initial filtrate production), then put back the ones to be kept (selective reabsorption), and throw away the unwanted ones (urine removal).

The Bowman's capsule gives way to a part of the winding renal tubule called the proximal tubule. The cells lining it are unlike those in any other area of the nephron. Instead of being flat and interlocking they have brush borders, with their outer membranes thrown up into thousands of fingerlike projections or microvilli. This greatly increases the surface area of this portion of tubule. (The small intestine also employs microvilli to increase its surface area, for absorption of nutrients.) And the presence in these cells of many mitochondria—the cellular "power plants" that supply energy—reveal that energy-consuming reactions are actively occurring here. In fact, 80 percent of the water filtered out

Kidneys

of the blood is reabsorbed by this portion of the nephron, along with 65 to 70 percent of the sodium that got through the gomerular filter.

Strangely, urea is also reabsorbed—although it is one of the principal unwanted materials in urine. Its re-entry into the body is an unavoidable side-effect of renal chemistry. About half of the urea passing into the tubule at the capsule end is withdrawn back into the body. But successive cycles from blood to glomerular filtrate make sure of an acceptable throughput of urea, with eventual expulsion.

Where do the reabsorbed substances go? Encircling the tubule are thousands of tiny blood vessels, ready to receive them via the wall of the tubule and their own walls, and transport them back through the venous system. The vessels are known as stellate veins because of their star shape. Blood from the glomerular capillaries flows into these tiny veins and becomes "rejuvenated" with the reclaimed water and salts. It then travels onward as the veins unite and grow, forming an enlarging network that heads toward the renal pelvis. At the pelvis, the veins come together to form the renal vein, which transports blood back to the heart.

IMPORTANT HORMONES
In addition to removing wastes, your kidneys release three important hormones: erythropoietin, renin, and the active form of vitamin D.

AROUND THE LOOP OF HENLE

Meanwhile, back at the nephron, the developing urine passes from the proximal tubule into the downward-sweeping loop of Henle. This delves into the kidney's medulla, becoming as it does so much thinner in diameter (down to five-thousandths of an inch) and with thinner walls. Metabolically, this descending portion of the loop is passive. Water passes through its wall to the blood vessels beyond by simple osmosis—the natural passage of water through a membrane from a weak solution to a more concentrated one, until both solutions are the same strength. In this way, the urine becomes more concentrated.

On the upward part of the loop, the cells in its wall are thicker and fatter, with no brush border. These cells are thought actively to pump glucose, sodium, and other desirable mineral salts, from the urine and through themselves, out to the surrounding blood vessels. This part of the reclamation process, like that in the proximal tubule, requires energy because it has to run against the prevailing concentration gradient.

One staggering feature of the nephron is that, amid the tangles of capillaries, tubules, and supporting cells within renal tissue, each nephron is virtually self-contained. The capillary of a glomerulus winds onward but sticks with its own tubule, encircling it intimately all the way around the loop of Henle and to its final portion, the distal tubule.

Kidneys

REGULATING SODIUM

Back in the cortex, at the "far end" of the loop of Henle, is the portion of convoluted tubule known as the distal tubule. This is involved in the regulation of sodium, a valuable body commodity which is used to adjust the balance between extracellular and intracellular fluids.

In the urine, sodium is in the form of positively charged ions (Na+), along with accompanying negative ions of chloride (C1−) and bicarbonate (HCO3−). Most of the sodium (up to 70 percent) and other ions that pass through the glomerular pressure filter, into the Bowman's capsule, is reabsorbed in the proximal tubule. The distal tubule reclaims less, but its action can be varied by the hormone aldosterone, to "fine-tune" the body's sodium balance.

The first stage in hormonal regulation of sodium reabsorption begins in the kidney. When flow through the renal blood vessels increases, the vessel walls release the substance renin. Once in the bloodstream, renin stimulates a vital biochemical transformation—the conversion of a plasma "prehormone" substance into the near-hormone angiotensin I, which then undergoes a further chemical conversion in the blood to angiotensin II. The hormone angiotensin II acts on the two adrenal glands, on top of the kidneys, and causes

them to release another hormone, aldosterone. (Angiotensin II also acts to raise blood pressure by increasing the resistance to blood flow.) The aldosterone is carried around in the blood and acts on the distal tubules of the kidneys to increase their rate of sodium reabsorption.

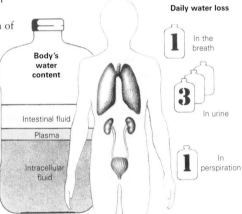

ABOVE More than half the human bodyweight is water, contained in the blood, lymph, cells, and intercellular spaces. Each day we drink an average of 5 pints (2.8 l) of fluid, and to keep the body in balance we also lose 5 pints (2.8 l) each day. 1 pint (0.57 l) is lost as water vapor in the breath, 1 pint (0.57 l) in perspiration, and 3 pints (1.7 l) in urine as water filtered from the bloodstream by the kidneys.

BALANCE
The kidneys help regulate the acid-base balance (the pH) of the blood and body fluids.

Kidneys

So more aldosterone saves sodium, while less allows its escape.

Sodium and the other ions cannot be reabsorbed alone, however: they are in solution and because of this they "drag" water with them, from the tubules into the blood vessels.

> **THREE IMPORTANT HORMONES**
> For regulating blood pressure—renin. Stimulation of the bone marrow to make red blood cells—erythropoietin. Maintaining calcium for bones and normal chemical balance in body—active vtamin D.

MAINTAINING WATER BALANCE

The distal tubules from different nephrons unite to form a combined or collecting tubule. Its wall gradually tapers, increasing in thickness as it unites with other collecting tubules and heads down to the renal pelvis.

Scientists believe that these collecting tubules show a unique response to antidiuretic hormone, ADH, one of the many bloodborne chemical messengers secreted by the pituitary gland at the base

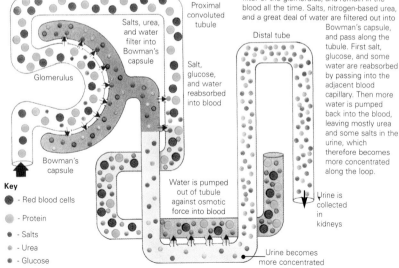

BELOW Different substances pass in and out of the blood at different stages of its journey around a kidney's nephron. Red blood cells and protein molecules are too big to pass through the filter of the glomerulus, and remain in the blood all the time. Salts, nitrogen-based urea, and a great deal of water are filtered out into Bowman's capsule, and pass along the tubule. First salt, glucose, and some water are reabsorbed by passing into the adjacent blood capillary. Then more water is pumped back into the blood, leaving mostly urea and some salts in the urine, which therefore becomes more concentrated along the loop.

Salts, urea, and water filter into Bowman's capsule

Proximal convoluted tubule

Distal tube

Glomerulus

Salt, glucose, and water reabsorbed into blood

Bowman's capsule

Key

- Red blood cells
- Protein
- Salts
- Urea
- Glucose

Water is pumped out of tubule against osmotic force into blood

Urine is collected in kidneys

Urine becomes more concentrated

Kidneys

of the brain. When the concentration of solutes in the extracellular fluid rises, as happens when the body is dehydrated, special cells in the brain detect the increased concentration and switch on the release of this hormone. ADH travels in the blood to the kidney, "instructing" it to reabsorb more water.

Under the influence of the higher blood level of ADH, the cube-shaped epithelial cells that line the collecting tubules become permeable and allow water to enter, possibly because the hormone opens up little pores in their membranes. This allows water to pass from within the tubule to the surrounding tissue, so continuing the concentration process of urine. If the urine in the tubules therefore remains dilute, little water is reabsorbed into the body. Depending on the prevailing internal conditions, urine can be four times as concentrated as the body fluids (it looks dark yellow when this happens), or if necessary up to three times as dilute.

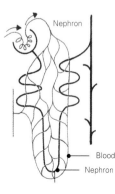

Nephron

Blood

Nephron

THE KIDNEYS AND BLOOD PRESSURE

The hormones ADH and aldosterone both have another important job to do—both are involved in the long-term maintenance of blood pressure. The link between blood pressure and urine production is clearly shown by the following sequence of events.

There is an accident. The victim suffers severe wounds and begins to bleed profusely; this means a sudden fall in blood volume, and a precipitous drop in blood pressure. The body's immediate response is to shut down many of the blood vessels, except those supplying the most vital organs such as the heart and brain, as well as increasing the heartbeat rate.

In the longer term, as blood clotting seals the wounds, the body needs to restore its fluid loss. Enter the kidneys. The low blood volume stimulates release of ADH, which makes sure the maximum amount of water is reabsorbed from the tubules back into the body's fluid systems. Meanwhile the low blood pressure initiates the release of renin, which, as described above, increases the angiotensin II level. This in turn increases the aldosterone level, which, via the distal

WHY DO KIDNEYS FAIL?
Most kidney diseases attack the nephrons, causing them to lose their filtering capacity.

Kidneys

tubules, pulls more sodium and other ions—as well as their accompanying solvent, water—from the urine into the body. This helps replace the fluid loss and restore the volume, and pressure, of the blood.

What happens when the situation is reversed and the plasma volume expands or blood pressure is raised? Renin release is inhibited, which means aldosterone secretion is reduced, and the release of ADH is minimal. The net outcome is that less sodium and water is reabsorbed, and the body rids itself of extra fluid. This helps reduce blood volume and lower blood pressure.

HYPERTENSION AND KIDNEYS

People with hypertension (persistently high blood pressure) may be put on a low-salt diet. The reasoning follows the link explained above between blood pressure and urine production. With low levels of salt in the body, very little renin and angiotensin II (which has blood pressure-raising effects) are free in the blood. Furthermore, if less salt needs to be reabsorbed by the kidneys, less water is reabsorbed too, which again helps to keep blood volume, and hence pressure, in check.

Many disorders of the kidneys, including either high or low blood pressure, can damage the nephrons and interfere with the kidney's filtering process. Normally nitrogen-containing wastes such as urea, from protein breakdown, are removed into the urine. An ill kidney cannot oblige, however. Protein-derived compounds such as urea build up in the body—and they are toxic. This is why people suffering from a kidney disorder may be put on a limited protein diet, to help keep down the level of urea in the blood. Proteins with a high "biological value," in eggs, milk, and lean meat, help increase the utilization of urea, thereby lowering its level in the blood. A high calorie intake is also needed so that the body does not turn to proteins for energy, which would break down to more nitrogenous wastes.

INTO THE RENAL PELVIS

It is time to return to the developing urine, flowing along the collecting tubules, eventually to leave the kidney and pass along the ureter to the bladder. The journey within the kidney is not quite over. Running up from the funnel-shaped end of the ureter and pelvis, to meet the medulla, are larger urine-collecting tubes or calyces, which branch and disperse into the body of the organ. The bulges of medulla between the calyces are known as papillae. One papilla collects urine from 70,000 nephrons.

DIABETES
Diabetes is a disease that keeps the body from using glucose (sugar) as it should. If glucose stays in your blood instead of breaking down, it can act like poison.

Kidneys

Just before reaching the top of each papilla, the collecting tubules bearing urine from the nephrons join together to form the ducts of Bellini. The papillae also contain the lower portions of the hairpin-shaped loops of Henle, before the tubules sweep back up to the cortex again.

As a result, this area of the kidney has two separate structures available to reabsorb water and concentrate the urine. It is known as the counter-current mechanism. If the renal papillae are not functioning correctly, urine leaves the tubules overdiluted and full of sodium. This vulnerable area of the kidney bears the brunt of any infection or damage from chemicals or drugs.

The collecting tubules, therefore, join up to form the renal papilla, which sits in a calyx in the renal pelvis. Waves of muscular contraction (peristalsis) coax the urine from the calyces, through the renal

pelvis and down the ureter. The muscles of the calyces are involuntary, and it is thought their contractions are triggered by distension in the tubules (a similar system occurs in the muscular wall of the heart). Once the peristaltic waves have been initiated, they travel 10 in. (25 cm) along the full length of the ureter from kidney to the bladder.

BELOW Most protein in the food we eat is broken down into amino acids during digestion, and stored or processed by the liver, which releases urea as a waste product. Some protein in the urine is normal, especially after exertion, but larger amounts are not. High blood pressure may cause swelling of tissue and release of protein, resulting in edema, and low blood pressure may cause an increase in plasma proteins. The excess may pass to the kidneys in the bloodstream and be filtered out, causing the presence of protein in the urine (proteinuria), which is usually symptomatic of metabolic disorder.

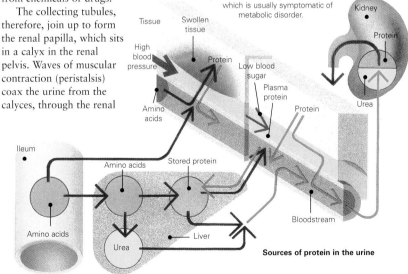

Tissue · Swollen tissue · High blood pressure · Protein · Low blood sugar · Plasma protein · Amino acids · Protein · Kidney · Protein · Urea · Ileum · Amino acids · Stored protein · Amino acids · Liver · Urea · Bloodstream

Sources of protein in the urine

Kidneys

THE STORAGE ORGAN

An empty bladder looks small and wrinkled, like a prune. This expandable bag is made from a basketwork of interlacing involuntary muscle fibers which run in whorls, sweeping up and down from the apex (top) to the neck (outlet) of the organ. Collectively, the fibers are known as the detrusor muscle. The detrusor's unique layout of fibers means that when the muscle contracts, the exit hole at the bladder's neck opens.

The two ureters track down the back of the abdomen and feed into the bladder via two small flaplike openings. Due to the angle at which the ureters enter the bladder, the openings—ureteric orifices—function as one-way valves. So, in health, when the bladder contracts to expel its contents, the urine cannot flow back up the ureters ("reflux") toward the kidneys. Developmental abnormalities or disease may result in faulty ureteric orifices into the bladder. They may need to be repaired by surgery, because the reflux of urine up to the kidneys can carry with it bacteria or viruses lurking in the bladder.

An average bladder can usually hold about ½ pint (0.28 l) of urine before it reminds its owner that it needs emptying.

> **"RENAL"**
> Anything that is related to the kidneys is called renal, for example, blood is carried into the kidneys by the renal artery.

If ignored, it can continue to expand in capacity to nearly 1 pint (0.57 l), but by this time thoughts are of little else except finding a lavatory. Yet the bladder can continue to stretch if necessary, as happens in certain disorders involving retention of urine. In men, for example, the prostate gland is located at the exit from the body. Enlargement of the prostate, or disease may swell and stiffen the gland, so that it squeezes shut the urethra and blocks the exit of urine. In some cases the bladder swells to accommodate up to 3½ pints (2 l) of urine.

The bladder's inner lining is formed from a folded, wrinkled membrane, known technically as the transitional cell epithelium. As the bladder expands, so the epithelium flattens out to maintain its integrity as a resistant, "urine-proof" barrier. However, a small inverted triangle of lining is unwrinkled. This is the trigone, which has at its two upper corners the ureteric orifices, and the urethral opening at its lower, central point.

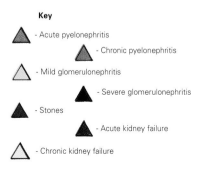

Key

- Acute pyelonephritis
- Chronic pyelonephritis
- Mild glomerulonephritis
- Severe glomerulonephritis
- Stones
- Acute kidney failure
- Chronic kidney failure

Kidneys

Poorly functioning kidneys can cause a host of varied symptoms affecting many different parts of the body. For instance, acute pyelonephritis—inflammation of the kidney substance and its pelvis— causes dark urine, back pain, nausea, and fever, whereas in its severe form glomerulonephritis causes fluid retention, resulting in low blood pressure, a reduction in urination, and ankles swollen by edema. Nausea and pain that moves from the upper to lower abdomen may be symptomatic of kidney stones. They may also cause dark or cloudy urine, or blood in the urine. Most serious of all is kidney failure, which in its chronic form causes a wide range of symptoms from

cramp and high blood pressure to thirst and increased urination at night; vision may also be affected. None of these symptoms should be ignored, and all need medical attention and treatment.

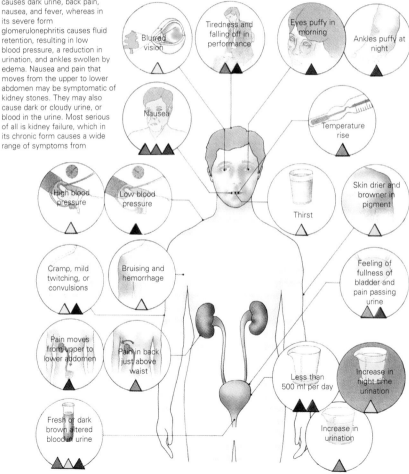

Blurred vision

Tiredness and falling off in performance

Eyes puffy in morning

Ankles puffy at night

Nausea

Temperature rise

High blood pressure

Low blood pressure

Thirst

Skin drier and browner in pigment

Cramp, mild twitching, or convulsions

Bruising and hemorrhage

Feeling of fullness of bladder and pain passing urine

Pain moves from upper to lower abdomen

Pain in back just above waist

Less than 500 ml per day

Increase in night time urination

Fresh or dark brown altered blood in urine

Increase in urination

Kidneys

INFECTION AND INFLAMMATION

Women are at an anatomical disadvantage as regards the risk of germs invading the urinary tract. Because the female's urethra is much shorter than the male's, it provides an easy route for infection from the exterior. Cystitis, inflammation of the bladder, may be due to invasion by hostile microbes or to irritant substances in the urine.

Inflammation of the kidneys themselves goes by a number of names, depending on which parts of the organ succumb. Glomerulonephritis affects the filtering mechanisms, is associated with various underlying disorders, and accounts for up to 30 percent of cases of chronic kidney (renal) failure. Infections by some strains of Streptococcus bacteria, which cause sore throats and scarlet fever, sometimes lead to glomerulonephritis, which occurs a few weeks after the initial infection. If glomerullar disease is severe enough to bring on sudden increase in blood pressure and hematuria, the condition is known as acute nephritis.

In acute pyelonephritis, the inflammation—often due to infection—is concentrated in the renal pelvis and tubules. Pregnancy, urinary obstruction,

KIDNEYS MAKE URINE
As each kidney makes urine, the urine slides down a long tube called the ureter and collects in a storage sac that holds the urine.

diabetes, and certain drugs are predisposing factors, while the symptoms include fever, hematuria, and shivering attacks (rigors). If the infection remains untreated, abscesses form in the kidney and permanently scar it.

Most patients with chronic pyelonephritis have obstruction and reflux, as well as possible infection. Unlike the acute infections, there may be no specific symptoms and the condition is revealed only during routine health screening, although some patients suffer from high blood pressure or general malaise. It accounts for 20 percent of cases of chronic renal failure.

FAILURE OF THE KIDNEYS

When the kidneys can no longer cope, for any of more than a dozen reasons, the result is renal failure. It may be acute (sudden) or chronic (coming on slowly). Acute renal failure develops over hours or days. Causes range from severe blood loss to severe glomerulonephritis, ingestion of a poison (such as antifreeze), to sudden and complete urinary obstruction. Provided medical help is obtained swiftly, possibly with intravenous feeding and short-term dialysis (with a "kidney machine") to tide the patient over the critical period, eventual recovery is almost certain. But it can be six weeks before the kidneys start producing urine again, and perhaps years before their function returns to normal—if it ever does.

Chronic renal failure is the major

Kidneys

cause of death from kidney disease. Its symptoms are many and varied, often vague and minor at first. They result from a build-up of metabolic wastes in the bloodstream, combined with breakdown of the barriers that prevent leakage of substances from blood into urine, and a failure of other essential processes controlled by the kidneys. Analysis of urine samples tells the clinician how far the disorder has progressed, through the appearance in them of protein, red and white blood cells, and "casts" (protein-derived material dumped into the tubules and washed down into the bladder). When the kidneys fail there are two chief forms of treatment. One involves a machine, the other, someone else's kidney.

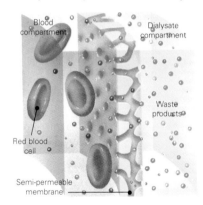

RIGHT Total kidney failure once led to death within a few days. Today patients with this condition—perhaps waiting for a kidney transplant—can be given a new lease of life with an artificial kidney machine. It uses the principle of selective filtering or dialysis to mimic the function of a real kidney and separate potentially toxic waste products from other components of the blood. The key part of the machine, the dialyser, consists essentially of a container of pure water in which dialysate concentrate has been dissolved. Blood from an artery in the patient's arm is pumped into the dialyser, where it passes through a network of tubes made from a semi-permeable membrane. This membrane has the property of holding back blood cells, plasma, proteins, and other large molecules, but it lets through urea and other waste products.

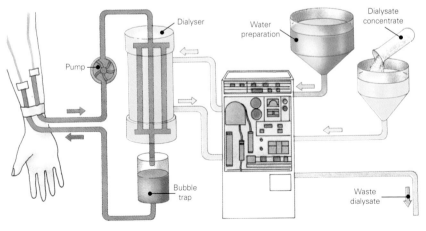

SENSES

Ceaselessly gathering information; your eyes, ears, tongue, nose, and sense of touch provide a rich flow of signals to the brain, which interprets them and gives reality to the objects that stimulate them. And what powerful and emotive sensations they can be! Aided by memory, the sound of a songbird and the smell of a rose can conjure up a pleasant image of a country garden. On the other hand, even just the sight of blood might make you faint or vomit.

Sight and hearing are generally regarded as the most vital of the senses—the loss of either is considered a real calamity, placing significant restrictions on any person's experiences and way of life. The senses of taste and smell are usually allocated an inferior role, though this perception grossly understates their true importance.

Sensors for the basic sensations of touch and temperature are distributed all over the body, so that you can feel a fur rug with your toes or test the temperature of the baby's bath water with your elbow. The more specialized sense organs for sight, hearing, taste, and smell are concentrated in the head. This comparatively exposed position is not without its risks, so most of the actual sensor cells are well protected by the bony structure of the skull—which also protects the all-important brain. Also, having the eyes, ears, nose, and mouth on the head brings with it the advantages of mobility, making it easy to direct sensory attention to anything of special interest—or toward any potential danger. It also allows short, fast, nervous connections between the sensors themselves and the controlling and analyzing brain.

In fact, the sense organs are, in reality, elaborate and highly sensitive extensions of the central nervous system. All of the sensations detected by them trigger minute electrical impulses which travel along direct nervous pathways to the brain. Once there they are coordinated and processed to give an ever changing mental image of the world around us, stimulating a wide variety of conscious and unconscious responses.

THE RICHNESS OF VISION

Sight is by far the richest of our senses, and it accounts for around three-quarters of our perceptions. With our eyes open, sensations come flooding in and are carried directly to the brain along the optic nerve. In principle the eye is like a camera: there is a lens system at the front to collect and focus light rays; the iris acts like a camera's aperture control; and the retina corresponds to the film which captures the images. There are even lens caps—the eyelids! But there is one big

THERE ARE FIVE
Each sense, and there are five of them: seeing, hearing, feeling, smelling, and tasting, has a specific purpose.

Senses

difference—unlike photographic film the retina can be used time and time again, continually capturing images at a rate of ten per second throughout its working life.

For safety the eyes are cradled deep inside bony bowls in the skull. Lining the eye sockets there is a fatty layer that cushions shocks and gives a highly lubricated surface for the continual swiveling of the globular eye. Six muscles direct this movement and anchor the eyeball securely in place.

Automatic responses provide further protection for the eye and the vulnerable associated nervous system. Sudden flashes of bright light, loud noises, rapid movements near the eye, and even a particle of grit trapped by the eyelashes trigger an immediate response as the eyelids slam shut. When closed the tough fibrous plates of the eyelids form a waterproof and airtight shield over each eye.

Sealing the eye opening from lid to lid, and lining the eyelids themselves, is the conjunctiva, a transparent membrane that catches anything that gets past the first line of defenses. It is kept moist by a thin layer of slightly oily tears produced by the lachrymal glands above the eye and spread by the blinking action of the eyelids. Irritation, or the presence of a foreign body in the eye, makes tears flow—a reaction that can also be stimulated by mirth or misery. Tears contain a mild antibacterial agent, lysozyme, which provides the eyes with additional protection against harmful bacteria in the air.

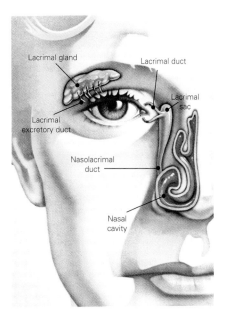

ABOVE Stimulated by sorrow, joy, or the emotionless accident of a piece of grit under the eyelid, tears are produced by the lacrimal gland. They flow over the surface of the eyeball, washing away germs and particles of dirt. Tears then pass along the lacrimal ducts and drain into the lacrimal sac, before draining into the nasal cavity by the nasolacrimal duct. When the eyes are watering copiously, the salty tears may be tasted in the throat.

PHOTORECEPTORS
There are two types of photoreceptors in the eye: rods and cones. The rods are responsible for vision in dim light conditions, the cones are responsible for color.

Senses

FOCUSING THE IMAGE

The white outer layer of the eyeball, the sclera, has a transparent circular segment that bulges out at the front to let light in. This is the cornea, which bends the incoming light and directs it toward the center of the eye. The bulging shape of the cornea is a precise curve which flattens out at the edges to minimize aberrations that would otherwise distort the image.

From the outside the sclera appears an opaque white, but its inner surface has a large number of blood vessels; this choroid layer is the main way in which essential nutrients are supplied to the eye. The choroid also contains a layer of the dark pigment melanin, which absorbs excess light. It also ensures that the interior of the eye remains dark, enabling the incoming light rays to give a good image on the retina—exactly as in a camera.

The cornea is also living tissue and to survive and work properly has to have a continuous supply of oxygen and nutrients. Other body tissues obtain these vital requirements through the blood. But

DIM LIGHT
The receptors in your eyes called rods, provide no information of color, it is the other photoreceptors in your eye—cones, that are used, but they do not work in dim light, so you cannot see colors in dim light. Check it out for yourself.

in the cornea, blood vessels would interfere with the passage of light. So instead it gets most of its oxygen directly from the air, absorbing it through the tears. Nourishment comes via the aqueous humor, a watery transparent fluid that fills the space behind the cornea.

CONTROLLING LIGHT INTENSITY

Immediately behind the cornea, the choroid layer takes on an additional function; it forms the iris, a colored muscular diaphragm that surrounds the black central opening of the pupil. This precise ring of muscle constantly contracts and expands, narrowing and enlarging the size of the pupil to control the amount of light entering the eye. In bright conditions it can narrow to as small as six-hundredths of an inch in diameter, while in the dark it may open up to as much as a third of an inch, gathering in whatever light is available. The action is fully automatic and triggered by the intensity of the light reaching the retina, but powerful emotions such as anger and fear, and certain drugs, can also affect the degree of iris opening.

The iris is the colored part of the eye, and the complete range of colors found in people's eyes is produced by different amounts of the same pigment—the melanin found in the rest of the choroid layer. Blue and green eyes have a minimal amount of melanin, with the amount increasing to give the other colors. Most

Senses

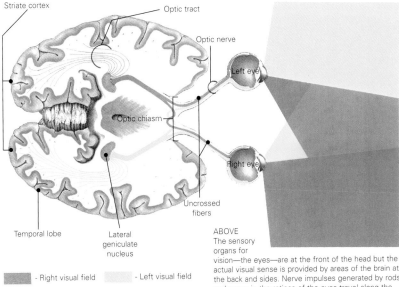

Striate cortex

Optic tract

Optic nerve

Left eye

Optic chiasm

Right eye

Uncrossed fibers

Temporal lobe

Lateral geniculate nucleus

- Right visual field   - Left visual field

ABOVE
The sensory organs for vision—the eyes—are at the front of the head but the actual visual sense is provided by areas of the brain at the back and sides. Nerve impulses generated by rods and cones in the retinas of the eyes travel along the optic nerves to the optic chiasma, where they partially cross over.

newborn babies have blue eyes because at first the melanin is concealed in the iris; after a few months it moves to the surface and the eyes take on their final color.

FINE IMAGE ADJUSTMENT

The pupil is the front surface of the eye's lens; it always looks black because it is an opening that leads into the dark center of the eye. Incoming light is bent toward the center by the cornea, and then finely focused by the lens to give a sharp image on the light-sensitive surface of the retina. With a camera, focusing is achieved by moving the lens elements backward or

forward, but the eye achieves the same result by the much more elegant process of changing the shape of the lens. This is the process of accommodation.

The lens is held in place by 70 fine ligaments—the ciliary zonule—which radiate from its edges like the gossamer strands that support a spider's web.

COLORS
A person's eyes can detect 10,000,000 different shades of color.

Senses

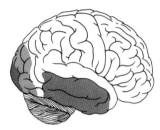

ABOVE Mixed impulses from both eyes pass via the optic tracts to the striate cortex at the back of the brain, before terminating in the temporal lobe vision area. In this way, right and left halves of the visual field merge. A blow to the back of the head jolts the striate cortex, sending false signals to the temporal lobes and making us "see stars."

The ligaments are, in their turn, attached to the choroid by the circular ciliary muscle. When the eye is at rest, the ciliary zonule pulls on the lens and flattens it. In this state the lens brings into sharp focus all objects that are more than 20 ft. (60 m) away.

For nearer objects the lens has to take up a more rounded shape, and to achieve this the ciliary muscle contracts inward, relieving the tension on the ligaments so that the lens bulges into a more spherical shape. The amount of accommodation is limited, and a normal eye is unable to focus much closer than 6 or 7 in. (15 or 18 cm) in front of the lens, a distance known as the eye's near point.

THE EYE IS SO SENSITIVE
The human eye is so sensitive that a person sitting on top of a hill on a moonless night could see a match being struck up to 50 miles (80 km) away!

The lens is made up of a series of layers, like those in an onion. Each layer refracts the light slightly to give a smooth and gradual overall effect. More layers are added throughout life, and as they accumulate the older ones pack together to form a hard core. This rigid core reduces the flexibility of the lens, so that the amount of accommodation gradually reduces and the near point moves farther from the eye as we grow older.

The back of the lens is around one third of the way to the retina; the light rays make the rest of their journey through the jellylike vitreous humor filling the inside of the eyeball. Like the air in a football, the vitreous humor keeps the eyeball under pressure and so maintains its spherical shape.

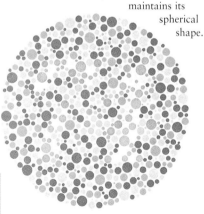

ABOVE One person in 20 is color blind. In the most common form, red-green color blindness, the retina lacks cones that detect red or green. Somebody without green cones has difficulty in seeing the number 74 among the dots in the above chart.

Senses

THE SECOND SENSE

If vision is the primary and most important human sense, second place must go to hearing. The sounds you hear, and make, play a vital role in communication through the reception of speech, or the pleasures of music, as well as warning you of impending danger. Virtually all sounds are made up of vibrations in the air, and the ears have the

BELOW A section through the outer and middle ears shows the eardrum and the three tiny ossicle bones (incus, malleus, and stapes). The inner ear, with its semicircular canals, has part of the spiral cochlea cut away to reveal the three channels inside.

task of detecting them and transforming them into electrical nerve impulses ready for the brain to analyze.

Sound vibrations correspond to pressure waves in the air, which spread out from their source like ripples on a pond. With simple sounds the vibrations have a regular waveform, rising smoothly

SOUND WAVES
Sound waves cause the eardrum to vibrate. Humans can hear sound waves with frequencies between 20 and 20,000 Hz.

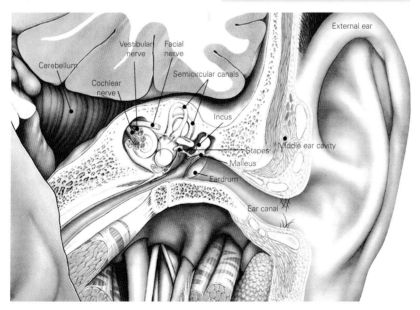

Senses

to a positive maximum, falling away to a negative minimum of the same magnitude, and rising again to the maximum in regularly repeated cycles. Complex sounds are merely built up from a series of superimposed simple waveforms.

The frequency of a sound is the number of cycles it completes in a second and is measured in hertz (Hz), where 1 Hz is one cycle per second. Frequency is recognized as the pitch of a sound; high-pitched notes have high frequencies and low tones have low frequencies. For humans the hearing range runs from about 20 Hz to 20,000 Hz, though the range varies a lot with age and among different individuals.

Loudness is just as important as pitch, and the ear is able to respond to an incredible loudness range, from the faintest whisper to the roar of a jet engine. Indeed the range is so large that scientists have to use a logarithmic scale to measure it, comparing any given sound to an agreed reference level. For practical purposes the unit used to measure sound levels is the decibel. Because the scale is logarithmic, a doubling in sound intensity is equal to an increase in 3 decibels. Thus a sound of 93 decibels is twice as loud as one of 90 decibels.

INSIDE THE EAR

Working out how sound waves are converted to electric nerve signals is simplified by the fact that the ear mechanism has three clearly identified and separate parts. These interconnected sections are the outer, middle, and inner ear, which each have their own specific functions, processing the sound in sequence.

The outer ear collects the sounds, which are transferred through the middle ear to the inner ear, where they are converted into nervous signals. The only noticeable component of this complex system is the fleshy external ear, known as the pinna, which leads into the external ear canal. Together they form the outer ear, which has the important functions of tuning in sounds and locating them. The outer ear increases the sensitivity of the

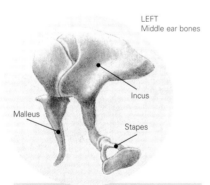

LEFT
Middle ear bones

Incus

Malleus

Stapes

THE STAPES
The stapes is the smallest bone in the human body. It is only 0.25 to 0.33 cm (0.10 to 0.13 in.) long and weighs only 1.9 to 4.3 mg.

Senses

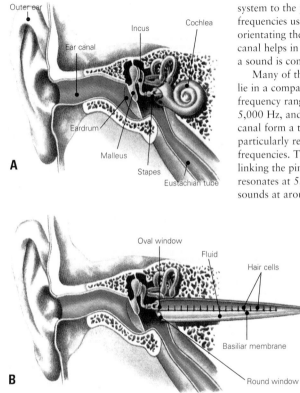

system to the particular sound frequencies used in speech, and orientating the trumpet-shaped ear canal helps in locating the direction a sound is coming from.

Many of the sounds of speech lie in a comparatively narrow frequency range of 2,000 to 5,000 Hz, and the pinna and ear canal form a tuned system that is particularly responsive to these frequencies. The conical opening linking the pinna to the ear canal resonates at 5,000 Hz, so that quiet sounds at around this frequency build up into larger vibrations. Similarly the ear canal itself resonates at around 2,500 Hz, and working together these effects amplify sounds that lie in the critical speech band.

ABOVE and OVERLEAF One of the main functions of the ear is to change sound waves into mechanical vibrations that can stimulate nerve cells. The key to the system are the three small ossicle bones, aptly named the malleus (hammer), incus (anvil), and stapes (stirrup). These bones are hinged together and fit inside the middle ear cavity (A).

For ease of explanation, the cochlea can be pictured as an uncoiled, tapering tube (B), with the basilar membrane and its rows of hair-cell receptors running down the center. Incoming sound waves are funneled along the ear canal (C), where they strike the eardrum and make it vibrate. The vibrations are picked up and amplified by the ossicles and passed on to the oval window in the cochlea. Vibrations of this window set up a traveling wave in the cochlear fluid, distorting the basiliar membrane and triggering the hair cells to send nerve impulses to the brain.

Senses

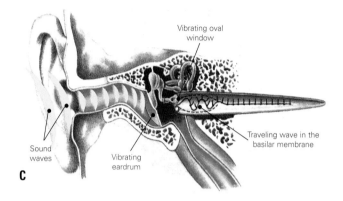

Vibrating oval
window

Traveling wave in the
basilar membrane

Sound
waves

Vibrating
eardrum

C

SOUND LOCATION

Identifying where the sound is coming
from is rather more difficult. It depends
largely on sensing differences in the
intensity and timing with which sounds
from a single source reach the two ears.
For example, a sound coming from the
left-hand side reaches the left-hand ear
before it gets to the right-hand one, and it
is also more intense at the left ear.

This discrimination does not work for
sounds coming from directly in front or
behind, and locating them relies on the
shape of the pinna. Some of the sound

waves from a source behind the head are
scattered by the edge of the pinna and
interfere with unscattered waves. The
result is a reduction in the intensities of
the frequencies in the range 3,000 to
6,000 Hz before they reach the eardrum.
These interference effects change as the
sound source moves around the head, and
the complex variations in frequency and
intensity can be analyzed by the brain to
give the required position information.
Similar effects are probably used to give
height information about sounds.

THE SMALLEST BONES IN THE
BODY

Sounds captured by the outer ear end up
at the eardrum, which lies at the end of
the ear canal and forms an airtight seal
between the outer and middle ears. The
middle ear is, however, still an air-filled
cavity; it is connected to the outside by

THE TEMPERATURE
At a temperature of 68°F (20°C), sound
travels at 1,125 ft/sec (343 m/sec). This is
the same as traveling at 756 m/hr (1,217
km/hr). Also, as the temperature rises,
the speed of sound gets faster.

Senses

the Eustachian tube. This is a narrow channel about one and a half inches long which runs from an opening at the back of the nasal cavity, to emerge through the floor of the middle ear. This connection ensures that the air pressure is the same on both sides of the eardrum—otherwise it would bulge under the pressure difference and so not be able to vibrate properly, or even burst. If the Eustachian tube becomes blocked—say by a head cold—the pressures become unequal and there is a temporary mild hearing loss.

Running across the cavity of the middle ear is a linked chain of the three smallest bones in the body, the auditory ossicles. Called the malleus, incus, and stapes, these tiny bones provide the connection between the eardrum and another membrane, the oval window, which forms the boundary with the inner ear. The snaillike coils of the inner ear are filled with fluid, and the task of the middle ear is to transform the air pressure changes caused by sounds in the outer ear into fluid pressure changes within the inner ear.

The middle ear mechanism is a miracle of living engineering. It is needed because the direct transfer of vibrations between air and fluid is very inefficient; in such a set-up, most of the vibrational energy would be reflected away from the fluid rather than absorbed by it. Using a mechanical linkage overcomes this problem and allows efficient energy transfer to take place.

DID YOU KNOW?
Why do you get dizzy if you spin around? When you spin, fluid in the semicircular canals of your ear moves around. This stimulates the hair cells. When you stop spinning, the fluid still moves a bit. Because the fluid is still activating hair cells, your brain still gets a message that you are moving, so you feel dizzy.

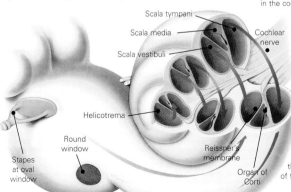

LEFT In ever decreasing circles, vibrations in the cochlear fluid (red arrows) spiral around channels to stimulate the organ of Corti to send nerve impulses to the brain. Initially the fluid vibrations are created as the oval window is vibrated by rapid movements of the baseplate of the stapes. On the outward trip, the vibrations travel along the scala vestibuli. Then, at the helicotrema (the apex of the cochlea), they change direction and spiral back (blue arrows) along the scala tympani. They finally arrive at the base of the cochlea, where they are dissipated by vibrations of the round window.

Scala tympani
Scala media
Scala vestibuli
Cochlear nerve
Helicotrema
Round window
Reissner's membrane
Stapes at oval window
Organ of Corti

Senses

A VERSATILE ORGAN

The tongue is one of the most versatile organs in the human body. Its extensive complement of muscles allows a range of complex movements for chewing, sucking, and swallowing, together with the vital function of modulating sounds to produce speech. But over and above these muscular functions it has a major role as the pre-eminent taste organ, and is also a highly sensitive touch sensor.

Most of the tongue consists of intimately interlaced muscles, arranged as paired blocks so that the left and right sides have independent sets of muscles. The two halves are divided by a fibrous septum running down the middle—externally this division is indicated by a central groove, or median furrow, running lengthwise along the tongue. Each block of muscles is formed from two arrays: a set of intrinsic muscles within the tongue for fine movements, and a set of extrinsic muscles connecting the tongue to the surrounding bones and used for large-scale movements.

Although it may seem to be floating freely in the floor of the mouth, your tongue is actually anchored in all directions by the four extrinsic

ABOVE The tongue is a flexible organ whose change of shape is brought about by the intrinsic muscles inside it. Gross movements of the whole tongue are controlled by the extrinsic muscles, which take up a surprisingly large amount of space within the lower jaw (mandible). There are sets of four muscles on each side of the head, which between them can pull the tongue in almost any direction.

DIFFERENT TASTES

According to current research, humans can detect five basic taste qualities. These are: salt, sour, sweet, bitter, and umami (the taste of monosodium glutamate and similar molecules).

Senses

muscle sets—the genio-glossus, hyo-glossus, stylo-glossus, and palato-glossus. The genio-glossus runs from the front of the lower jaw into the tongue from tip to base. Contraction of these muscles (on either side) makes the tongue stick out as its whole foundation is pulled forward. A flat straplike muscle, the hypo-glossus passes from the side of the tongue down to one arm of the wishbone-shaped hyoid bone in the throat. Movement of these muscles pulls the sides of the tongue downward.

Linking the sides of the tongue to the base of the skull through the bony styloid process are the stylo-glossals. They act to pull the tongue backward and upward. Connected to the sides and rear of the tongue the palato-glossals run to the rear of the palate and lift the sides of the tongue when they are contracted.

Working together these extrinsic muscles have the flexibility to move the tongue in virtually any direction. But the movements they produce are relatively coarse, and fine shape changes are the province of the intrinsic tongue muscles. Again these are arranged in four groups: two run from the front to the back of the tongue; one runs transversely; and the fourth runs vertically.

HOW WE TASTE
Flavor molecules fit into receptors on the microvilli at the top of the cell, causing electrical changes that release transmitters on to the nerve ending at the bottom of the cell.

RECEPTORS FOR TASTE

The actual taste buds are microscopic organs—no more than 20 to 40 millionths of an inch wide—containing 30 to 80 receptor cells, many of which are connected to the endings of separate nerve fibers. The buds are buried in the surface layers of the papillae and connect to the outside through a narrow duct known as the apical pore.

Detailed examination using an electron microscope shows that the taste buds have a complex structure. The body of the bud contains no less than three different cell types, known as Types 1, 2, and 3. All three types extend from the base of the bud upward to end in fingerlike projections in the apical pore.

The precise function of each cell type has not been firmly established, but it is believed that cells of Types 2 and 3 are the actual taste receptors, while Type 1 cells have a supporting function. All are renewed on a regular basis, surviving for only about ten days before being replaced. A fourth cell type, basal cells, are also found in the taste bud and these may be newly formed replacements that are about to develop into Type 1, 2, and 3 cells.

Taste receptor cells do not have their own nerve axons, instead they make contact with the ends of sensory nerve fibers running in the body of the tongue. Indeed in an embryo the taste buds form under the influence of the sensory nerve fibers. This close relationship is

Senses

maintained in an adult; if the connection between nerve and bud is broken, the bud rapidly degenerates. But if the nerve regrows into the surface layer of the tongue, it stimulates the growth of new taste buds.

PROTECTION
Your sense of taste protects you from unsafe foods. If you ate poisonous or rotten foods, you would probably spit them out immediately, because they usually taste revolting. That way, you stop them from entering your stomach.

TASTE BUD DISTRIBUTION

The papillae on the upper surface of the tongue carry a high concentration of taste buds, but this is not the only part of the mouth where taste buds are found. There is a thin scattering in many other parts of the mouth's mucous membrane, including the epiglottis (a small cartilage flap that helps seal the windpipe, or trachea), the larynx (voice box), the soft palate, and

BELOW Resembling mushrooms in a field of grass, fungiform papillae protrude through the filiform papillae on the surface of the tongue. Like the other types of papillae further back, they bear hundreds of taste buds.

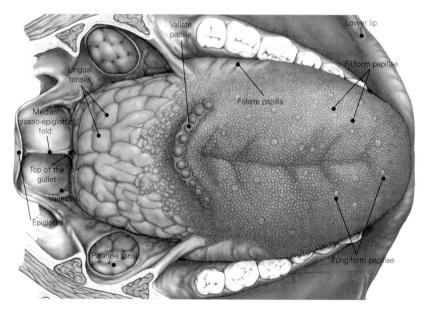

Vallate papilla

Lower lip

Lingual tonsils

Filiform papillae

Foliate papilla

Median glosso-epiglottic fold

Top of the gullet

Vallecula

Epiglottis

Palatine tonsil

Fungiform papillae

Senses

the uvula (at the back of the mouth). There are even taste buds on the mucus membrane lining the upper third of the esophagus which means that you carry on tasting food as you swallow it.

TYPES OF TASTE

Everybody has favorite foods and favorite flavors, yet the taste buds themselves can recognize only five basic tastes: sweet, salt, sour, bitter, and umami. These characteristics are registered by different types of papillae on the tongue. Fungiform papillae respond to sweet and sour tastes, the vallate region responds best to sour and bitter sensations, and the foliate receptors also favor sourness.

A CHEMICAL BALANCE
Your sense of taste also helps you to maintain a consistent chemical balance in your body. Liking sugar and salt for example, satisfies your body's need for carbohydrates and minerals. Similarly, eating sour foods such as oranges and lemons supplies your body with essential vitamins.

BELOW In a cross-section of the central part of the tongue's surface, three kinds of papillae reveal their shapes. Most numerous are the conical filiform papillae, whose taste buds are sensitive to sour flavors. The buds of the flat-topped fungiform papillae respond to both sweet and sour tastes, whereas those of the vallate papillae—concentrated in the "moat" surrounding its central mound—react to bitter flavors. The gaps between papillae are kept moist with mucus secreted from glands at the base of the gaps. Flavor molecules must dissolve for the taste buds to detect them.

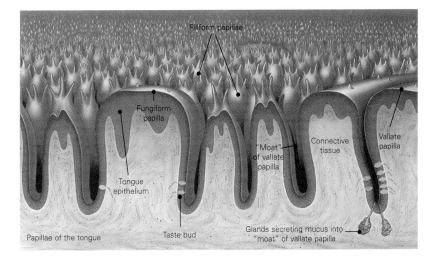

Filiform papillae
Fungiform papilla
"Moat" of vallate papilla
Connective tissue
Vallate papilla
Tongue epithelium
Papillae of the tongue
Taste bud
Glands secreting mucus into "moat" of vallate papilla

Senses

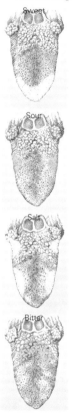

Transferring these characteristics on to the surface of the tongue gives a taste map; it shows that the tip has peak sensitivity to sugars, the sides to acid tastes, and the back of the palatine section to bitterness. This arrangement explains why saccharine, and similar sweetish substances, leave a bitter aftertaste. Initially the tip of the tongue responds to the sweetness, but the bitter taste is then detected as the sensors at the back of the tongue come into action.

LEFT Four of the five main basic tastes can be detected in different areas of the tongue as shown here, for instance sweet flavors are detected by the tip of the tongue.

FROM TONGUE TO BRAIN

Nerve fibers for the taste buds on the tongue come from the seventh and ninth cranial nerves. Cranial nerve seven—the facial nerve—supplies the front regions of the tongue and so handles sensations that are mainly related to salt and sweet tastes from the fungiform papillae. Bitter and sour sensations come from vallate and foliate papillae on the rear part of the tongue, and are dealt with by the ninth cranial nerve. Most of the other taste buds scattered throughout the mouth cavity and throat are supplied by the tenth cranial nerve, the vagus.

Most children like sweet foods whereas many of their elders prefer savory dishes. This difference may be a primitive hangover from the way in which taste sensations are processed in the brain. Signals from the taste receptors enter the central nervous system along two different routes, one to the subcortical regions of the brain and one to the cerebral cortex. Subcortical connections run to areas such as the medulla and cerebellum, which are some of the older parts of the brain in evolutionary terms. In particular the sensations route to the amygdala and hypothalamus regions, which appear to be responsible for nonconscious, or instinctive, reactions. A typical example is the way most people instinctively respond favorably to sweet tastes but reject bitter ones—unless they have consciously trained their tastes to more sophisticated responses, which favor bitter or savory sensations.

RECEPTOR CELLS
Each of your taste buds contains 50–100 specialized receptor cells. Sticking out of every single one of these receptor cells is a tiny taste hair that checks out the food chemicals in your saliva.

Senses

SOMETHING IN THE AIR

Smell is often considered to be the least important of our senses, but it may well be one of the oldest and probably acts more directly on our subconscious than the other senses. There is little doubt that scents have important roles in human behavior. Indeed the body is provided with glands to produce specific odors. Many of these appear to be associated with sexual attraction and excitement—they generally start to operate during puberty—but there are others that have considerable significance. For example, the bond between baby and mother is thought to be enhanced by a form of scent imprinting. A baby suckling at the breast pushes his or her face into a bank of scent organs that surround the nipple. Equally, the smell of a baby is almost universally regarded as pleasant, providing an added source of pleasure for the mother when she cuddles her child.

More overtly a whole section of our culture is devoted to the carefully calculated stimulation of the sense of smell—and hopefully of other responses such as sexual attraction—with carefully blended perfumes and scents. And a further indication of the powerful potential of the sense of smell is the way it can become a major source of information in the absence of other more immediate senses, especially sight.

OLFACTION
The sense of smell, called olfaction, involves the detection and perception of chemicals floating in the air. Chemical molecules enter the nose and dissolve in mucus within a membrane called the olfactory epithelium.

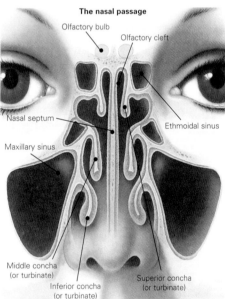

The nasal passage

Olfactory bulb

Olfactory cleft

Nasal septum

Ethmoidal sinus

Maxillary sinus

Middle concha (or turbinate)

Inferior concha (or turbinate)

Superior concha (or turbinate)

ABOVE Convolutions of cartilage and bone form a labyrinth of cavities within the nasal area of the face. The superior conchae, or turbinate bones, and olfactory clefts are lined with olfactory membrane. Sensory hairs in the mucous-covered membrane send signals upward to the olfactory bulbs, located behind the bridge of the nose. Each bulb has a direct connection to the brain along the olfactory nerves.

Senses

NASAL PASSAGES

Only a small part of the nose and nasal cavity (the internal space bounded by the floor of the cranium and the roof of the mouth) is taken up by the organs of smell—the rest of it is mainly concerned with processing the airflow on its way through to the lungs. The walls of the nasal cavity, and particularly the flaplike middle and inferior conchae, are coated with respiratory mucous membranes. These incorporate a vast number of tiny hairlike cells which act together to move sequential waves of mucus toward the throat. Dust, bacteria, and chemical particles inhaled with the air are trapped by the mucus, carried back, and swallowed; they are then dealt with by the gastric juices, which quickly nullify any potential harm.

The sense organs themselves consist of two yellowish-gray patches of tissue, the olfactory membranes, each about the size of a postage stamp. They lie in a pair of clefts located just under the bridge of the nose and at the top of the nasal cavity.

The yellowish hue of the olfactory membrane is due to the presence of two main kinds of pigment: carotenoids and

free vitamin A, and phospholipids. The reasons for the coloration are not altogether clear, but the pigments seem to be necessary for the membrane to work. Certainly albino animals (ones which have no pigments at all are known to have no sense of smell.

During normal breathing most of the air flows smoothly through the nose, with only a small fraction reaching the olfactory clefts—though this is enough to trigger a response to a new smell. Deliberate attempts to detect smells by sniffing make the air move through the nose much faster, increasing the flow that makes its way up to the olfactory clefts and so carrying more odor molecules to the sensors.

If you have a cold the production of respiratory mucus is several times greater than normal—so you have a runny nose. The mucus may obstruct the narrow openings to the olfactory tissue. This prevents odor-carrying air from reaching the sensory membranes and your sense of smell is restricted.

COLDS

As you probably know, when you have a cold and your nose is stuffed up, you cannot smell very well. This is because the molecules that carry smell cannot reach the olfactory receptors.

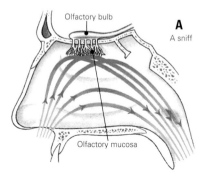

Olfactory bulb

A
A sniff

Olfactory mucosa

Senses

INSIDE THE OLFACTORY MEMBRANE

Viewed through an electron microscope the active surface of the nasal membrane looks rather like a plate of spaghetti in a sticky sauce. The spaghetti strands are olfactory cilia, or hairs, growing out from receptor cells, and the sauce is mucus produced by special glands (Bowman's glands) in the membrane. Each receptor cell—and there may be as many as ten million of them—ends in a tiny swelling, or olfactory knob, which has about five olfactory hairs growing out of it. The hairs may be as long as one-hundredth of an inch but are normally much shorter. They lie tangled together in the mucus and may have slight swellings where they

touch. Although the precise mechanism has not yet been fully established, the hairs appear to contain receptor sites that take up specific odor molecules and so trigger the receptor cell into producing a nerve impulse.

Just like the taste cells, smell receptor cells have only a short working life—they function for about a month and then start to break down and are removed by the mucous stream. Replacement cells are constantly being produced: they grow from cells in the base of the olfactory membrane, thrusting outward and developing olfactory hairs. One explanation for this short life is that the cells gradually become clogged up with

BELOW The flow of air through the nasal cavity varies during sniffing, breathing in, and breathing out. If you sniff hard to deliberately savor a smell (A), incoming air is forced up into the olfactory clefts to bombard the hairs of the olfactory membrane with odor molecules. During normal breathing in (B), little of the incoming air reaches the odor detectors. Breathing out (C) causes turbulent air flow, though odor perception is still maintained.

DIFFERENT ODORS
People can distinguish between 3,000 and 10,000 different odors. Some are more easy to detect than others, people are very sensitive to the smell of a green bell pepper, detecting it when it is mixed with air at only 0.5 parts per trillion.

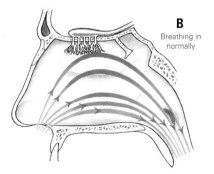

B
Breathing in
normally

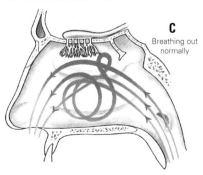

C
Breathing out
normally

Senses

absorbed odor molecules, but this concept is far from being universally accepted. Another possible reason is that the receptors are in a particularly exposed position for what are really sensitive nerve

cells, and so are more likely to suffer damage. With senses such as vision and hearing the receptor cells are much better protected and do not normally deteriorate.

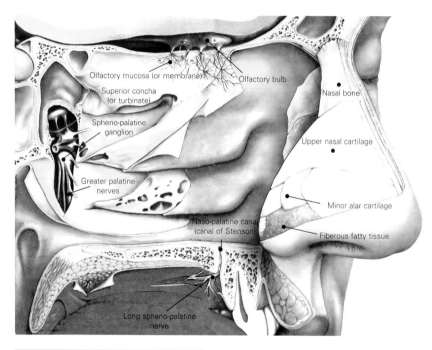

Olfactory mucosa (or membrane)

Superior concha (or turbinate)

Spheno-palatine ganglion

Greater palatine nerves

Olfactory bulb

Nasal bone

Upper nasal cartilage

Minor alar cartilage

Naso-palatine canal (canal of Stenson)

Fiberous fatty tissue

Long spheno-palatine nerve

ANOSMIA

About two million people in the United States have no sense of smell. This is a disorder called anosmia. A serious head injury can cause anosmia.

ABOVE The anatomy of the nose is shown in this cutaway view of the right side of the face. The central partition, or septum, between the nostrils has been removed to reveal the extent of the olfactory membrane and conchae (turbinate) bones.

Senses

WHY DO SMELLS SMELL?

Many theories have been proposed to explain just what gives a particular substance its specific odor. The most probable explanation is that the effect depends on the shape of the substance's molecules, and the way they lock on to (or into) the receptor sites on the olfactory hairs. Support for this theory is given by the way continued exposure to an odor rapidly results in odor fatigue so that the smell is no longer perceived. This fatigue may come about because the receptor sites are fully occupied, and a short break is needed before the sites are ready for further use and the specific smell sense can be reactivated.

THE BULB THAT SMELLS

The olfactory hairs are effectively a direct extension of the brain. The receptor cells they spring from run from the olfactory mucosa directly into a part of the brain called the olfactory bulb. The bulb is very small, about the size of a match head, and located slightly below the bridge of the

nose, about four-tenths of an inch into the head. It is separated from the olfactory membrane by the cribriform plate, a wafer-thin section of bone at the front of the cranial cavity.

Nerve fibers from the receptor cells pass through openings in the cribriform plate and run to synapses (gaps between nerve connections) in the bulb. These are concentrated in structures known as glomeruli, with the ten million or so receptor cells connecting to just 2,000 glomeruli. Two outputs run from each glomerulus: one set of 24 tufted cells connects to the olfactory bulb on the other side of the head, and another 24 cells—the so-called second olfactory neurons—run into deeper regions of the brain.

Each nerve cell in each group of 24 running from a glomerulus is like an electrical switch: it can be either on or off. Used like the binary code in a digital computer this arrangement allows a total of around 16 million permutations, giving plenty of scope for the identification of individual odors—if this really is the way that the odor signals are coded.

SCENT OR BOUQUET

The scent or bouquet of a wine is as much a measure of its qualities as its taste. Only when the smell has been thoroughly appreciated will the wine taster put the glass to his lips. The "nose" of the wine taster is so highly developed that he can distinguish between many wines by their smell alone.

RECEPTOR CELLS

Olfactory receptor cells live for only about a month, and function as odor-detectors for only a small part of that time before they break up and die. They develop from basal cells, which are continually produced to maintain the supply.

The
CYCLE OF LIFE

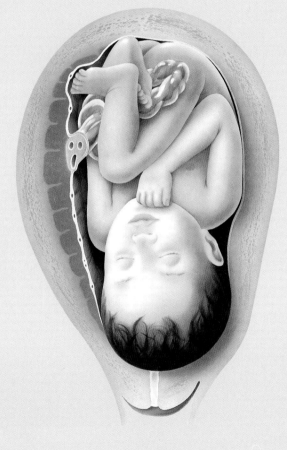

Human beings are arguably the most complex organisms on this planet. They are all built to the same basic plan, yet all are different—everybody is an individual. In many countries, people have an average lifespan of 70 plus. During the years between birth and death they undergo a remarkable and continual process of physical, emotional, and intellectual development that begins in the womb and continues until life's end. The way in which human beings grow and develop has been seen as a miracle which still fascinates philosophers, psychologists, and educationalists alike. Its study is also a major branch of medical science.

REPRODUCTION

The creation of human life is a wonder. Two minuscule cells—a male sperm and a female ovum—fuse together to form one new cell, so small that it is barely visible to the human eye. Within four days this cell implants inside a woman's womb where it rests, nourished by the mother. It grows and develops at a phenomenal rate until it emerges, nine months later, as a fully-formed, breathing, feeling new human being, a tiny replica of the two adults who gave it form.

GROWTH & DEVELOPMENT

Even today, despite the theories of Sigmund Freud, the Gessels, Jean Piaget, and others, there are still many unanswered questions. Scientists can chart human growth and development quite accurately, but still they do not know exactly why certain changes take place, or what are the major influences that lead to the creation of an individual person with unique characteristics.

GENETICS & HEREDITARY

In the whole complex and fascinating story of the human body, the reason why we are what we are is probably the most intriguing aspect of all. At conception, two sex cells—a sperm from a man and an ovum from a woman—fuse to produce a single fertilized cell. Within this one cell is contained all the information required to produce a new human being, and which determines every aspect of that human being from hair color and height to the essentials of artistic ability and intellect. This information is inherited from the previous parents, stored in the genes—thousands of which make up the threadlike chromosomes present in the cell—and carried in code on the double-helix molecule DNA (deoxyribonucleic acid) which in turn makes up the genes themselves.

REPRODUCTION

Since earliest times the wonder of conception and birth has fascinated thinkers, poets, dreamers, and philosophers who have woven myth and fantasy about the event. But reproduction, though fantastic, is not haphazard. Instead it calls into play all the most sophisticated workings of the human body, relying for its achievement on timing, specialized anatomical structures, and the concerted efforts of glands and hormones.

Like yin and yang, reproduction involves the fusing into one whole of two halves— male and female. Even with today's advances in medical science that have enabled such developments as test-tube babies, the male and female contributions toward the creation of a new human being remain basically the same. The male body is designed to allow the manufacture, storage, and transfer of male sex cells, or sperm. The woman's body is designed to manufacture, store, and release eggs, or ova. Together the two bodies provide an anatomical structure that dependably unites sperm and eggs.

THE MALE REPRODUCTION SYSTEM

The sexual organs of a man are partly visible and partly hidden within the body. Visible parts are the penis and scrotum, a saclike pouch suspended below the penis which houses the two egg-shaped testicles, or testes. Hidden inside the body, and also part of the male reproductive system, are the prostate gland, seminal vesicles, and a number of tubes such as the vas deferens, or sperm duct, which links the whole system together.

The penis transfers sperm to the woman's body during sexual intercourse. Normally flaccid, but becoming erect during sexual excitement, the penis consists of spongy tissue loosely covered with skin. Erection results when extra blood is pumped into the spongy tissue during sexual arousal. The rounded head of the penis, or glans, is separated from the shaft by a rim of tissue called the corona, or crown. The foreskin, a retractable hood of skin, normally covers the glans but is often circumcised at birth, a procedure now believed to lessen the risk of penile cancer. A tube along the center of the penis, the urethra, serves to carry urine from the bladder. But at the climax of intercourse it functions as part of the sperm duct and carries semen.

THE HUMAN SEX ORGANS
For a man these are: penis, testicles, prostate, seminal vesicles, epididymis, Cowper's glands.
For a woman these are: vulva (notably the clitoris), vagina (notably the cervix), uterus, fallopian tubes, ovaries, Skene's glands, Bartholin's glands.

Reproduction

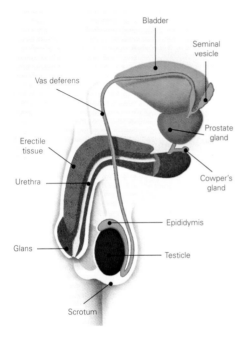

Bladder

Seminal vesicle

Vas deferens

Prostate gland

Erectile tissue

Cowper's gland

Urethra

Epididymis

Glans

Testicle

Scrotum

ABOVE The male sexual apparatus is partly inside a man's body and partly outside. Sperm are manufactured in the testes, or testicles, which are kept at the correct temperature hanging in the external scrotum. From there they travel to be stored in the epididymis. Normally the penis is limp but during sexual arousal blood pumped into its spongy tissue causes it to become erect. As the moment of ejaculation approaches, Cowper's gland, the seminal vesicles, and the prostate contribute fluids that mix with the sperm to lower its acidity and form semen, which muscular contractions squirt along the vas deferens and out of the urethra.

THE SPERM INDUSTRY

Sperm production takes place inside the testicles, at an astonishing rate. During a lifetime, from puberty onward, one man may produce as many as 12 trillion sperm. Each sperm takes about 72 days to mature, and the entire operation is controlled by a complex interaction of hormones, and made possible by the thermostatlike quality of the scrotum, which keeps sperm at the correct temperature.

Sperm are manufactured from parent cells called spermatogonia lining the seminiferous tubules, which are coiled tubes within the testicles. When a spermatogonium divides, one half remains behind for future division and the other half migrates into a layer of specialized Sertoli cells, which secrete fluid and nourish the new cells.

HORMONES

The reproductive cycle in both men and women is regulated by several different hormones. Some of these include testosterone, estrogen, and progesterone.

Reproduction

At this stage the young sperm cell, like others in the body, contains its full complement of 46 chromosomes. It divides further to produce four spermatids, each containing only 23 chromosomes.

Within the Sertoli cells the young sperm acquire the characteristic tadpole shape of mature sperm. There are two seminal vesicles, in which the sperm cells are stored until fully mature; the vesicles also supply the fluid part of semen. Within the epididymis and vas deferens sperm remain fertile for some weeks.

HORMONE CONTROL

A complex interplay of hormones triggers and controls sperm production. It begins at puberty as a boy matures into a man. The testes produce not only sperm but also testosterone, the male sex hormone responsible for the development of male characteristics such as hair on the face and body, the deepening of the voice, and the enlargement of the external sexual organs. The trigger for the teenage growth spurt in boys, testosterone also brings about its end, halting the growth of the long bones in the limbs at about the age of twenty.

Testosterone itself is secreted in response to chemical commands within the brain. Sparked off by these, the hypothalamus releases a hormone called gonadotropin-releasing factor. This stimulates the pituitary gland to release two hormones—luteinizing hormone (LH)

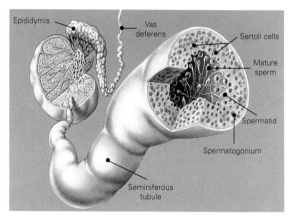

Epididymis

Vas deferens

Sertoli cells

Mature sperm

Spermatid

Spermatogonium

Seminiferous tubule

TESTOSTERONE

This is produced by the testes in the male and stimulates the development of male secondary sex characteristics, like facial hair and a deeper voice.

ABOVE The sperm factory consists of thousands of seminiferous tubules coiled in the testes. Within their edges, Sertoli cells nourish embryonic germ cells which gravitate toward the center of the tubule to become spermatids and then full-fledged sperm. The mature sperm—at a rate of up to 500 million every day from both testes—are held in store in the epididymis. When called for, they travel along the vas deferens to the penis.

Reproduction

and follicle-stimulating hormone (FSH). These enter the bloodstream and are carried to the testes, which begin to produce testosterone. Hormone levels are kept constant by an ingenious feedback system whereby rising testosterone levels suppress production of gonadotropin-releasing factor, which in turn reduces the release of LH by the pituitary. As testosterone levels gradually drop, so releasing factor and LH are stimulated, which again get to work on the testes and make them produce more sex hormone.

THE FEMALE REPRODUCTION SYSTEM

The term "equal but different" so often applied to women and men is particularly apt when it comes to reproduction. For while both play an equal part in conception, it is the difference which enables successful reproduction. Whereas the man produces and transfers sperm, it is the woman's body which produces, stores, and releases the vital other half, the ova. In addition it is the woman's body which receives sperm, shelters the developing fetus, and finally gives birth to the new infant.

ABOVE RIGHT The testes respond to and produce hormones. Sperm production by the testes is stimulated by hormones originating in and near the brain. The hypothalamus produces gonadotropin-releasing factor to make the nearby pituitary gland produce luteinizing hormones (LH) and follicle-stimulating hormone (FSH). These travel in the bloodstream to the testes and induce them to manufacture sperm and testosterone, which gives male characteristics.

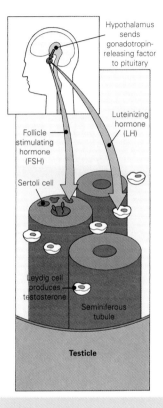

Hypothalamus sends gonadotropin-releasing factor to pituitary

Luteinizing hormone (LH)

Follicle stimulating hormone (FSH)

Sertoli cell

Leydig cell produces testosterone

Seminiferous tubule

Testicle

ESTROGEN
This hormone is produced by the ovaries in the female and stimulates the development of female secondary sex characteristics (wider hips and mammary glands) as well as starting the thickening of the uterus lining in preparation for a possible pregnancy after the egg is released.

Reproduction

Unlike those of a man, the woman's sexual and reproduction organs are almost entirely hidden, so that in a standing position the only visible sign is the pubic hair. A woman's sexual organs are known collectively as the vulva. They consist of a mound of fatty tissue, the mons pubis, which, from puberty, is covered with pubic hair. Extending from the mons are two folds of tissue, the outer lips, or labia majora, which protect the reproductive and urinary openings between them. Between them lie the labia minora, or inner lips, delicate, hairless and sensitive folds of skin. Below the mons they split into two folds to form a hood under which lies a small budlike organ, the clitoris. Descended from the same embryonic structure as the penis, the clitoris too swells during sexual arousal and is the most sensitive part of the woman's sexual organs.

Lying just below the clitoris is the exit of the urethra (which leads to the bladder) and the vaginal opening, the external entrance to the vagina. A muscular tube about 4–6 in. (10–15 cm) in length, the vagina is capable of great distension. It provides a port of entry for the penis and sperm as well as extending considerably during labor to allow a baby to be born. A thin membrane, the hymen, seals the vagina in virginity, but is usually ruptured during the first sexual intercourse.

The upper end of the vagina leads into the cervix, the neck of the womb or uterus. A muscular pear-shaped organ which expands to a hundred times its normal size during pregnancy, the uterus lies in the lower abdomen. Extending outward and back from each side of the uterus are the Fallopian tubes, which extend their feathery ends toward the two ovaries.

The ovaries, like the male testes, have a dual function. They produce the female sex hormone, estrogen, and house the female sex cells, or ova. Unlike a man, however, a woman does not manufacture sex cells. Instead the newborn female is born with full complement. Known in their immature state as cocytes, they number about 600,000 at birth, and are contained in small saclike primordial follicles within the ovaries. In theory each one has the potential to ripen and be fertilized; in practice only about 400 ova ripen during a woman's lifetime.

THE MENSTRUAL CYCLE

A woman's reproductive years begin with puberty and end with the menopause. Controlled by the hypothalamus in the brain, major changes occur from about the age of eight. The hypothalamus begins

PROGESTERONE
This is produced by a yellow tissue called corpus luteum in the empty ovarian follicle—this hormone maintains the thickness of the uterus lining in case fertilization and development of a fetus occurs.

Reproduction

to secrete gonadotropin-releasing substance. This acts on the pituitary gland, causing it to release various hormones. Follicle-stimulating hormone (FSH) is the first to be released. This brings about the growth of the egg-containing follicles and makes them produce estrogen, which encourages the growth of breasts, widens the pelvis, and encourages the development of the external genitals.

The rising level of estrogen in the bloodstream has an effect on the hypothalamus known as "negative feedback," which causes a reduction of a second substance, luteinizing hormone (LH). This causes one of the follicles to burst and release an ovum for possible fertilization, the escape being known as ovulation. Ovulation frequently passes unnoticed but may be accompanied by a mild cramp or mitteldsmerz.

The remains of the follicle, now known as the corpus luteum, stay in the ovary secreting estrogen and a second hormone, progesterone, which prepares the uterus lining (the endometrium) to receive and nourish a fertilized egg. If the egg is not fertilized, levels of estrogen and progesterone drop and the uterus lining breaks down. It is then shed from the vagina together with mucus and the unfertilized ovum. The resulting bleeding constitutes the menarche, or first menstruation.

This same menstrual cycle then repeats itself roughly every 28 days throughout a woman's reproductive life. It begins around the age of 11 and ends with the menopause at about the age of 50, when the ovaries become exhausted and can no longer manufacture estrogen and progesterone. The menstrual cycle first becomes irregular and then ceases altogether, a change that is often accompanied by other symptoms such as hot flushes, depression, vaginal atrophy, and so on.

CONCEPTION

Walt Whitman has described sexual drive as "the procreant act of nature" and certainly sexual intercourse, while providing considerable sexual pleasure, is also nature's most effective way of transferring sperm from the male body to the female. Four stages of sexual arousal have been defined—excitement, plateau, arousal, and orgasm—each of them accompanied by distinctive physical changes. In men the first stage begins with erection of the penis brought about when the spongy tissue becomes engorged with blood. In women arousal is also accompanied by changes in the external genitals and lubrication of the vagina.

THE MALE ORGANS
Unlike the female sex organs, the male ones are mainly outside the abdomen. Hanging below the main body, the testes are usually a degree or two cooler than the main body temperature.

Reproduction

The penis penetrates the vagina which expands to accommodate it. Intensity builds and muscular contractions of the testicles, epididymis, and vas deferens force sperm into the urethra. On their way sperm mix with seminal fluid secreted by the prostate gland and seminal vesicles to form semen which, at orgasm, is ejaculated into the woman's vagina.

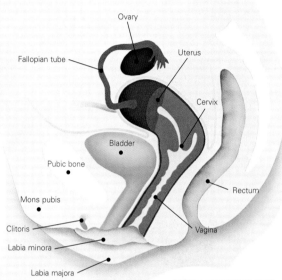

In one ejaculation a man may send as many as 500 million sperm into a woman's vagina, although only one will fertilize an ovum. Once in the vagina, sperm swim upward by lashing their tails, aided by prostaglandris which dissolve the mucus plug at the entry to the uterus.

Many sperm fall by the wayside, but tens of thousands arrive in the uterus. There a glucose-rich environment helps them to move along the Fallopian tubes where, if the woman has ovulated within the previous 48 hours, one mature ovum will be waiting. The first sperm to reach the egg penetrates its surface, releasing an enzyme called hyaluronidase. This slices through the chemical coating of the ovum

ABOVE and OPPOSITE PAGE
Female sexual organs are located entirely within a woman's body. They consist of two ovaries, situated near the feathery entrances of the Fallopian tubes, which lead to each side of the uterus, or womb. The cervix—the neck of the womb—leads in turn to the vagina, whose exterior opening is between the clefts of the liplike labia. The ovaries have two principal functions: to produce and periodically release ova, or eggs, and to manufacture the female sex hormones estrogen and progesterone. The vagina is the sheath that holds the man's penis during intercourse and receives ejaculated sperm.

SPERM CELL

The average sperm has a head containing the genetic material, a middle piece containing lots of the energy-converting parts called mitochondria, and a long, whippy tail.

Reproduction

and the two cells fuse. At the same time the ovum surface closes and enzymes are released which dislodge other sperm and create an impenetrable barrier. With this action the ovum is fertilized.

Twins occur either if the ovum, after fertilization, splits into two (in which case the babies will be identical twins), or if two ova have been released and fertilized.

In this latter case each ovum is fertilized by a different sperm, and as a result the mother gives birth to non-identical or fraternal twins.

The sex of the future member of the human race is decided at the moment of conception. Both sperm and ovum contain only 23 DNA-carrying chromosomes, each half the number in all other body cells. The ovum's 23rd chromosome is either an X-chromosome or a Y-chromosome. If an X-sperm fuses with the ovum, the resulting embryo is female; if a Y-sperm fuses it is male. In this way, fertilization provides the new cell with its full complement of 46 pairs of chromosomes—all the genetic material needed to sustain life and determine the development of the future human being. It is at this point, too, that the genetic defects such as Turner's syndrome may be passed on.

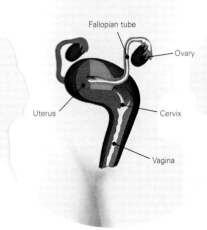

Fallopian tube

Ovary

Uterus

Cervix

Vagina

THE WHOLE SPERM
The whole sperm is only 0.002 in. (50 micrometers) long, and most of its length is the tail.

IMPLANTATION

Within just a few hours the fertilized egg, now known as a zygote, begins to grow and develop, moving down from the Fallopian tube propelled by a fingerlike cilia. Cell division begins almost immediately as coils of deoxyribonucleic acid (DNA) carried on chromosomes reproduce themselves. The fertilized egg divides into two new cells, then four, then eight, and so on. By the fourth day the zygote, now comprising 16 cells, enter the

Reproduction

uterus. As it continues to divide, a fluid-filled amniotic cavity develops within the structure, now known as blastocyst. Some cells flatten out to form an outer mass, the trophoblast (from which the placenta will subsequently develop); the remainder cluster at one end to form the embryoblast (from which the embryo will develop).

Small projections, chorionic villi, form on the trophoblast and, about one week after fertilization, burrow into the uterine wall. The blood vessels in the villi tap into the maternal bloodstream across a band of tissue ultimately to become the umbilical cord. This chorionic villi also produce a third hormone, human chorionic gonadotropin (HCG). Tests for the presence of this hormone in a woman's bloodstream give a reliable indication of pregnancy. HCG also maintains the corpus luteum in the ovary, so that it goes on producing estrogen and progesterone, which are responsible for the signs of early pregnancy such as breast tenderness, nausea, giddiness, and the cessation of ovulation and menstruation.

Cell differentiation takes place with each cell division—a subtle but nevertheless crucial aspect of development that can have a seemingly dispro-portionate repercussion such as a stunted limb—perhaps resulting from the loss of or damage to just one single cell. The embryoblast divides into two layers of cells, an outer endoderm and an inner ectoderm. Together they form the embryonic disk. During the third week this disk grows and lengthens, and a fissure develops along the length of the ectoderm. The ectoderm's cells slip toward this and spread sideways to form a third layer of cells, the mesoderm, sandwiched between the other two. The amniotic cavity also expands and encloses the entire embryonic disk and its connecting body stalk in a fluid-filled sac, within which the growing embryo has freedom of movement and is cushioned from the outside world.

THE EMBRYO

Over the next month the embryo undergoes dramatic changes. By the middle of the third week, it measures about 1/8 in. (0.3 cm). Primordial, budlike forerunners of organs and tissues begin to appear and cells move through the mesoderm to create a rod which eventually becomes the spine. The nervous system also develops first as a neural rod, but then its upper end swells to form the brain.

Chunks of mesoderm called somites form in pairs on each side of the neural

EGG CELL
A typical egg cell is about 0.005 in. (120 micrometers) across. It has the mother's genetic material in its nucleus, and a hazy layer around it called the zona pellucida.

Reproduction

tube. From these begin to emerge muscle, skin, and skeletal tissue. Heart and blood vessels also arise from the mesoderm, the heart beginning life as two tiny tubes which subsequently fuse into a single chamber.

By the end of the first month the primitive heart pumps blood around the embryo. During the second month sensory tissues, central, and peripheral nervous systems develop from the ectoderm. Bone cartilage, muscle, and essential organs such as the liver, kidneys, and spleen form from the mesoderm, while the endoderm gives rise to the lining of various tubes in the body: the digestive, excretory, and respiratory tracts.

Arms and legs begin to appear by the end of the first month, first as primitive flippers, then developing hands, arms, shoulders, and legs. Head growth is rapid—at this stage the embryo measures approximately half an inch, half of which is head. Within two more weeks the heart, now a series of chambers, is pumping blood, the brain continues to develop, and the respiratory system expands. At seven weeks, the embryonic skeleton begins to turn from cartilage into bone, and by the tenth week facial features—eyes, nose, ears, and chin—have appeared.

Initially fishlike in appearance and life-maintained through the placenta, by the end of the second month the embryo has taken on distinctly human features.

BELOW An ovary. Gonadotropin-releasing factor from the hypothalamus triggers the pituitary to produce follicle-stimulating hormone (FSH), which commands the follicles in the ovary to make estrogen.

Discharged ovum

Mature corpus luteum

Mature follicle

Fallopian tube

Ovum

Ovary

Developing follicles

THE EMBRYO

The fertilized egg divides, or cleaves into two cells, then four, eight, and so on. The result is the blastocyst, or ball of cells. At about four weeks the embryo is about as big as a grape.

Reproduction

THE LIFE-SUPPORT SYSTEM

The developing embryo is nourished through the placenta, a disk-shaped organ which develops during the first ten weeks of life at the point where the chorionic villi first burrowed into the endometrium. The placenta, which is about 7 in. (18 cm) across and about 1 lb. (0.45 kg) in weight, has an outer (maternal) surface divided into lobes of chorionic villi and tiny blood vessels. The flow of blood to and from these vessels is supplied from the mother's uterine artery and vein. The inner surface of the placenta is covered by a layer of amnion, and has a series of tiny vessels radiating out from the umbilical cord at its center. The umbilical cord, which links the embryo to the placenta, houses blood vessels which carry blood to and from the embryo.

The placenta acts as both pool and filter. Cells on the maternal surface fill with blood, and from this pool blood vessels on the embryonic surface draw not only oxygen but also, by diffusion, proteins, vitamins, and other nutrients necessary for growth. Some, particularly

proteins, are also manufactured by the placenta itself. At the same time waste products are drawn from the embryo's blood vessels. There are, however, a number of drugs and other harmful substances that, if taken by the mother, can cross the placenta and cause damage to the embryo.

The placenta also replaces the work of the corpus luteum, producing estrogen, progesterone, and other hormones on which successful pregnancy depends.

0 hours Fertilization

30 hours 2-cell stage

3 days 8-cell stage

4 days 64-cell stage

5-6 days Balstocyst

Fallopian tube

Ovary

ABOVE and OPPOSITE PAGE The first month after conception sees miraculous changes as a fertilized ovum develops into an embryo implanted in its mother's womb. During the first few hours, the ovum divides repeatedly as it makes its way along a fallopian tube, arriving about six days later as a blastocyst that embeds itself into the lining of the womb. Over the next week or so this hollow ball of cells evolves various layers and cavities until there is a separate embryo and placenta. By about four weeks the embryo has a heart, rudimentary eyes, and buds from which the limbs will grow. By this time blood vessels within the umbilical cord supply the embryo with oxygenated blood from the placenta and carry waste products back to it.

THE FEMALE ORGANS
The uterus, or womb, is about the size and shape of a pear, near the base of the abdomen. It is linked to the outside by the vagina, or birth canal.

Reproduction

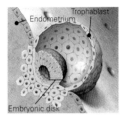

6–7 days
- Endometrium
- Trophoblast
- Embryonic disk

12–13 days
- Amniotic cavity
- Trophoblast

FETAL DEVELOPMENT

At eight weeks the embryo is a fully-formed minute infant, now called a fetus. Over the next 36 weeks growth takes precedence over development as the weight of the fetus increases 600 times from less than 1 oz. (0.45 kg) to around 7 lb. (3.18 kg), and it reaches its full length of 20 in. (50.8 cm). Bodily proportions change, too, as limbs and trunk grow, reducing head proportions to one-fourth of the body length.

Hair, eyebrows, and eyelashes are added during the 20th week, followed by fingernails and toenails. A fine, downy hair called lanugo covers the limbs and trunk.

Within its fluid-filled protection, the fetus also begins to move and swallow, taking in amniotic fluid. Thumb-sucking occurs during the fifth month, and one month later the fetus acquires the grasping reflex characteristic of a newborn infant. From 20 weeks the fetal heartbeat can be heard through a stethoscope. By 24 weeks the fetus can survive outside the womb if protected in an intensive care unit.

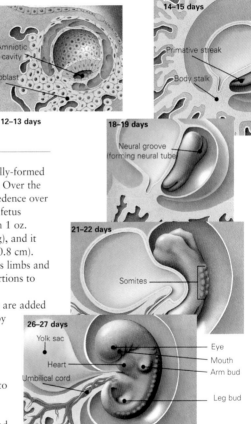

14–15 days
- Primative streak
- Body stalk

18–19 days
- Neural groove (forming neural tube)

21–22 days
- Somites

26–27 days
- Yolk sac
- Heart
- Umbilical cord
- Eye
- Mouth
- Arm bud
- Leg bud

AFTER TWO MONTHS

After two months or so, the baby is known as a fetus. It's main activity is to grow bigger, and add details such as eyelids, fingernails, and toenails.

Reproduction

THE CHANGING BODY

While the fetus develops inside the mother, her body also undergoes the dramatic changes of pregnancy. For the first three months there are few outward signs of what is happening; it is not until the fetus begins its growth spurt that a woman's abdomen becomes increasingly swollen. But other changes have been taking place from the moment of conception, most of them controlled by the hormones estrogen and progesterone. The action of progesterone inhibits ovulation and so menstruation—a missed menstrual period is often the first sign of pregnancy. Other symptoms include nausea, giddiness, swollen or tender breasts, and frequent urination. As pregnancy progresses estrogen causes the softening and enlargement of the pubic joints, ligaments, and tissues in preparation for labor. This action brings its own discomforts such as frequent urination, backache, and stretch marks as connective tissue loosens.

Hormones also affect the blood circulation. Blood volume rises by nearly 50 percent, to compensate for loss during childbirth, but estrogen increases the risk of blood clotting or thrombosis. Varicose veins and hemorrhoids are other possible unpleasant side-effects.

THE BABY
A baby grows inside its mother's womb for nine months before it is born.

The pregnant woman visits her obstetrician regularly throughout pregnancy, with the number of visits increasing during the eighth and ninth months. She receives a full physical examination and blood and urine tests. Blood is monitored for signs of anemia, quite common in pregnancy, and urine is checked for any signs of gestational diabetes, another condition peculiar to pregnancy. Later in pregnancy doctors watch carefully for signs of raised blood pressure that may indicate one of the most serious disorders—pre-eclampsia or eclampsia. Once a major cause of maternal mortality, eclampsia can now easily be detected and brought under control.

Fetal heartbeat is also carefully monitored, and the size of the fetus checked, most usually by ultrasound. By and large, however, the mother herself, by eating a balanced diet and following an exercise routine, can ensure a healthy pregnancy and a speedy return to her pre-pregnant shape. Later in pregnancy, too, most expectant mothers are encouraged to attend childbirth classes and learn the breathing routines so helpful in labor.

If a miscarriage occurs it is most likely to do so in the first three months of pregnancy. Causes are still uncertain but may include failure of the placenta to form, bacterial infections and hormonal dysfunction, or chromosomal abnormality. Always a distressing experience, miscarriages are usually a sign that the fetus was defective, and after a first miscarriage the chances of a second,

Reproduction

successful pregnancy are high. Throughout pregnancy fetus and mother are closely interlocked, so that external dangers can be passed easily to the fetus, causing harm. Chief among these are viral infections, venereal diseases, smoking, alcohol, and drugs. The most hazardous viral disease is rubella (German measles), which, if contracted by the mother during the first month of pregnancy, can cause serious fetal deformity. Research now indicates clearly that smoking during pregnancy can cause miscarriage, fetal death, premature delivery, and respiratory problems. Alcohol too can cause low birth weight, serious birth defects, and deformities. The disabling effects of thalidomide during the early 1960s also highlighted the dangers of drugs which, if taken by the mother, can cross over the placenta with devastating effects.

Birth defects are probably a woman's greatest fear, but the advancement of medical technology does mean that a number of chromosomal or genetic abnormalities such as Down's Syndrome or spina bifida can be detected before birth. Amniocentesis, for instance, whereby a sample of amniotic fluid is removed and analyzed, can be carried out at 15 weeks. It is gradually being replaced by chorionic villus biopsy, which can be carried out between the ninth and eleventh week. Chorionic villi contain a rich source of fetal DNA, so analysis of a small sample of chorionic tissue indicates possible genetic abnormalities.

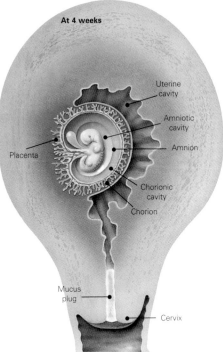

At 4 weeks

Uterine cavity

Amniotic cavity

Amnion

Placenta

Chorionic cavity

Chorion

Mucus plug

Cervix

THE PLACENTA
Inside the uterus, the baby is surrounded by fluid. It cannot breathe for itself or eat. It obtains oxygen and nourishment from its mother through the placenta embedded in the wall of the womb.

Reproduction

Embryonic development continues until at about four months the baby, now known as a fetus, is recognizably human. By seven months the fetus is fully developed and capable of independent, if precarious, existence outside the womb. During the last two months of pregnancy the fetus grows in stature and puts on weight.

6 weeks
½ in. (1.2 cm) long

8 weeks
1 in. (2.5 cm) long

9 weeks
2 in. (5 cm) long

4 months
7 in. (18 cm) long

ULTRASOUND
An ultrasound scan is usually taken from between 15–20 weeks. Harmless, very high-pitched sound waves, undetectable by our ears, are beamed through the mother's body. The resulting image of the baby in the uterus can be used to check the baby's heart, brain, and other organs —and to see if it's a boy or girl.

CHILDBIRTH

The birth of a baby can be the most thrilling and most traumatic experience for both infant and mother. Fashion, cultures, historical time, and even attitudes have shaped the method of childbirth in strange and wonderful ways. But as an event it is unique, and what seems evident today is that the more a woman is able to participate in the birth, the more joyous the experience.

Full-term pregnancy lasts nine months; it is followed by labor, during which the new infant is expelled from the mother's uterus. Throughout pregnancy the body prepares for labor with gentle contractions. During the last month these contractions gradually increase, culminating in labor. Quite what triggers labor is still uncertain. Some doctors believe that once the uterus is fully extended, others that decreasing levels of progesterone initiate the event. Oxytocin, a hormone secreted by the pituitary, is also known to stimulate the uterus, while prostaglandins may also play a part. Recent work also suggests that the fetus itself may contribute to the onset of labor.

Whatever the actual cause, labor itself is heralded by various signs. In first pregnancies the fetus settles head down about two to three weeks before labor, the cervix softens, the lower part of the uterus expands and the cervix effaces—pulls up

Reproduction

7 months

toward the body of the uterus—and dilates. As the cervix widens, a mucus plug, which has formed in the cervix to keep the uterus germ-free, slips out in what is known as the "show." It may be slightly tinged with blood. The amniotic sac slips down and ruptures, releasing the amniotic fluid, and labor begins.

Labor falls into three distinct stages. The first stage is the longest, an average of 14 hours for a first birth but less for second and subsequent births. It is characterized by persistent, regular contractions as the uterine muscles shorten and pull upward, pushing the fetus downward, and pulling the cervix upward. As the first stage progresses, contractions become increasingly intense, occurring every minute or two. The second stage begins once the cervix is fully dilated, to about 4 in (10 cm). This stage may take anything from a few minutes to two hours. During it contractions slow down and the baby begins its journey along the birth canal, the mother working with the contractions to push the baby down. In a straightforward birth, the head emerges first, the shoulders turn, and further contractions push the baby out into the world. The baby's mouth and nasal passages are cleared of mucus, the baby breathes, takes its first cry and is given to the mother. Once the umbilical cord has finished pulsating, it is clamped and cut.

Ninety-five percent of all births are head-first in the so-called vertex position. But in just over 3 percent of births, the baby is in a breech position, with buttocks or feet presenting. In some limited instances it may be possible to turn the baby; otherwise Caesarean section—delivery of the fetus through an incision in the mother's abdomen—is the most usual procedure.

In the third and final stage of labor, the uterus continues to contract, leading to the painless expulsion of the placenta and umbilical cord. The mother is checked carefully to ensure that the entire placenta has been removed, and with this the process of pregnancy and birth is complete.

As soon as the baby is born, a nurse in the delivery room assesses its physical condition by assigning an Apgar score of zero to two for each of five tests.

GROWTH & DEVELOPMENT

In particular some controversy still surrounds the fundamental question of how far any person is the product of the environment and how much simply the result of the mingling of two sets of genes. What is certain, however, is that with the act of conception a train of developmental events is set in motion which continues uninterrupted until death.

GROWTH BEFORE BIRTH

Human development in the nine months before birth is faster than at any other time after. Human life begins with conception when a male sperm fuses with a female ovum to produce a single cell. Over the next 40 weeks this single fertilized cell divides again and again, millions of times, in a process that transforms it from a minuscule speck no larger than a pinhead into a fully-formed miniature human individual completely equipped for independent life outside the womb.

The mechanism by which this occurs is called mitotic cell division, a copying and splitting by which cells constantly divide and reproduce themselves. This process, which begins in the womb, continues throughout life. It is controlled by genes, minute particles within the cells. Each gene carries within it a "blueprint" of information. Together, the millions of genes determine the unique physical and mental characteristics inherited by the new human being from its parents.

OUT OF THE WOMB

A full-term pregnancy lasts 40 weeks, or 288 days, although birth can be expected anytime from the end of the 38th week or day 266—the birth day is calculated by the obstetrician from the onset of the last menstrual period. Labor is thought to be triggered off by various hormones, including oxytocin, which is produced by the pituitary gland, prostaglandins, and catecholamines.

The effect of birth on a baby's growth and development is still a matter of medical speculation. Some experts, such as the French obstetrician Frederick Leboyer, believe that birth can be a violent and traumatic experience for the baby, who leaves the peace of the womb to enter what Leboyer describes as an "overwhelming confusing world." Leboyer, who recommends a highly relaxed "birth without violence" approach, is particularly opposed to the technologically based practices of the modern delivery room. Among these he includes Caesarean section and Apgar

PREGNANCY

The course of pregnancy is broken up into three sections, known as trimesters. During this time, the fetus develops from a single-celled zygote into a fully developed fetus.

Growth & Development

assessment, the range of tests carried out on a newborn baby within the first five minutes of birth. Advocates of high-tech delivery, however, argue that such developments have either saved lives of infants or, in the case of Apgar testing, have alerted physicians to babies' needs for immediate care.

An average full-term newborn weighs between 6–9 lb. (2.7–4.08 kg) and is about 20 in. (50.8 cm) long; the head accounts for about one-quarter of this length.

The skin is dry, soft, and wrinkled, and may still be covered with a soft down or lanugo. The eyes are almost always blue, although they may change color in later months, and the infant's vision is blurred. Hearing, too, may be slightly impaired because of amniotic fluid in the ears.

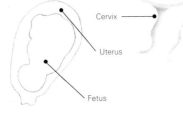

LEFT Within the sheltered domain of its mother's womb, a fetus undergoes nine months of miraculous development. Safe inside the amniotic sac and bathed in amniotic fluid, the fetus receives oxygen and nourishment extracted from the mother's blood by the placenta. They travel along blood vessels in the umbilical cord, which also contains vessels that carry waste products back to the placenta.

Placenta

Umbilical cord

Amniotic sac

Amniotic fluid

Cervix

Uterus

Fetus

EMBRYO

The embryonic stage is from conception through to the end of the eighth week. During this time, huge developments take place, transforming the mass of cells of the blastocyst into the embryo with distinct human characteristics.

Growth & Development

BORN SURVIVOR

Survival is the keynote to the first few days of life. The new infant, who has been protected in the warm, liquid environment of the womb for nine months relying on the life-support system of the placenta, has to make an immediate adjustment to the outside air-breathing world. Respiration is the first concern. At birth the lungs are still partly filled with amniotic fluid. During labor the fetal adrenal glands release the hormone epinephrine (adrenaline) to absorb the fluid so that it can be replaced by air. As a result the lungs should inflate as soon as the baby is born. Sometimes this fails to happen, perhaps because the umbilical cord is wrapped around the windpipe. Even so, the newborn is surprisingly resilient and can survive for short periods even with an inadequate supply of oxygen.

The circulatory system switches from total dependence on the placenta to becoming an independently functioning unit, a process that involves the closure of various pathways which, in the womb, allowed the fetus to receive oxygen without using the lungs. The kidneys are also immature, but are helped through the first critical days by hormones which maintain the correct balance of body fluids until the kidneys can take complete control of this function.

Finally, the newborn baby also has to adjust to digesting milk rather than receiving nutrients directly into the bloodstream via the placenta. Again the mechanisms for adjustment are highly efficient. In a breast-fed baby the first feed of colostrum, the yellowish fluid which precedes breast milk, encourages the release of hormones which, in turn, stimulate the gastrointestinal system, and in this way enable digestion and absorption to take place properly.

A newborn baby appears completely helpless. With a large head on a scrawny neck that cannot support it, small legs, and an uncoordinated nervous system, the baby seems unlikely to be capable of achieving an independent existence. But the survival instinct and the drive toward achievement are very strong indeed. During the first two years, growth and development proceed at a remarkable pace.

The timing of the various developmental stages varies, with different children arriving at key stages at slightly different ages. Nevertheless by their second birthday most children have traveled a long way from babyhood. By that time most can walk, talk, feed themselves, solve simple problems—in fact they are now, in many ways, mini adults.

THE PLACENTA
The placenta will provide nutrients and oxygen to the developing fetus for its entire time in the womb. The placenta acts as an intermediary between maternal and fetal blood, without allowing the two circulations to come into direct contact.

Growth & Development

SURVIVAL SYSTEMS

Babies need constant loving care in order
to thrive and develop both physically and
emotionally. The newborn baby arrives in
the world with a deeply built-in survival
system of reflexes and behavior.

Crying is a baby's first means of
communication. A cry may mean hunger,
tiredness, or discomfort, but it is a sound
that few parents can ignore. If food is not
what is needed, the parent picks up the
baby and provides comfort and love,
forming an emotional bond by which the
infant can thrive and to which he or she
increasingly responds.

Together with crying, the newborn baby
has other reflexes, some of which date back
to the very earliest days of human
existence. The sucking reflex is extremely
strong. When the baby's cheek is touched,
he or she automatically turns toward the
touch and the lips pucker. In this way when
the lips touch the breast or a feeding nipple,
the baby immediately begins to suck, so
ensuring successful feeding.

BONE DEVELOPMENT
Bone growth begins very early in fetal
development. At six weeks,
chondrocytes in the center of the
cartilage begin to enlarge—the first stage
in bone development. By the time baby is
born, bone has been laid down in the
center of the shaft and in a collar around
the middle of the shaft.

Shaft

Epiphyseal
cartilage

Epiphysis

Fatty marrow
with blood
supply

Growing
cartilage

ABOVE Ossification, the process of bone formation,
occurs in the long bones of the fetus at the shaft and
ends of cartilage "models." Cartilage continues to
form in the epiphyses, gaps between the ossification
centers, as the bone grows. Capillaries bring blood to
bone-forming cells, or osteoblasts, and fatty marrow
forms in the bone's hollow center. The marrow starts
manufacturing blood cells.

Growth & Development

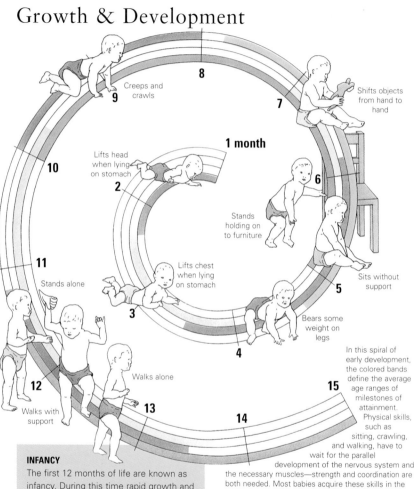

1 month

Lifts head when lying on stomach

2

Lifts chest when lying on stomach

3

Bears some weight on legs

4

Sits without support

5

Stands holding on to furniture

6

Shifts objects from hand to hand

7

8

Creeps and crawls

9

10

Stands alone

11

Walks with support

12

Walks alone

13

14

15

In this spiral of early development, the colored bands define the average age ranges of milestones of attainment. Physical skills, such as sitting, crawling, and walking, have to wait for the parallel development of the nervous system and the necessary muscles—strength and coordination are both needed. Most babies acquire these skills in the order shown, although some infants reach each stage before others. The beginnings and ends of the colored bands pinpoint the ages at which 25 percent and 90 percent of infants have attained a particular stage of development. The ages at which 50 percent have mastered a skill are indicated by the positions of the babies within the colored bands.

INFANCY

The first 12 months of life are known as infancy. During this time rapid growth and development can be observed, as baby's senses heighten, weight is gained, teething begins, the first steps taken, and first words are spoken.

Growth & Development

The grasp reflex causes a baby automatically to clench the fist if an object such as a finger is placed in the palm. This reflex disappears after the first three months, but is so strong that a baby can support its entire weight for a few seconds. The startle reflex consists of rapid movements. If startled or handled too briskly babies will throw back their heads and fling out their arms and legs. The movements are usually followed by crying. Most of these reflexes disappear as voluntary actions take over from this instinctive behavior.

In the first few weeks after birth, weight gain is the most usual way of checking that growth and development is on course. A baby's weight at birth may be affected in many ways. Boys, for instance, tend to be slightly heavier than girls; first babies are often slightly lighter than subsequent infants. Heredity also plays a part. Large parents usually produce babies who are larger than average, whereas small parents have smaller babies. Like many other human characteristics, such as color of eyes or hair, height, too, is genetically controlled.

CHILDHOOD

This is the period between infancy and puberty, and though rapid growth of infancy is not matched, this period of time sees significant physical and mental development and growth.

This control is exercised by polygenes, groups of genes which are passed down as blocks through the generations. Environmental factors such as smoking during pregnancy or maternal illness may also affect birthweight.

Most babies lose weight in the very first few days, but birthweight is usually regained within ten days. Thereafter the baby's weight should steadily increase in a consistent fashion.

ON THE MOVE

Newborn babies have little or no control over physical movement. They lie in the birth position, on the stomach, knees bent underneath, buttocks in the air. To the anxiety of many first-time parents, the baby flops helplessly if lifted unsupported. By six weeks, however, the baby has gained some control of the head, and as the muscles gain in strength the head gradually becomes less heavy in relation to the body. By seven to eight weeks most babies are strong enough to keep their bodies in line with the head when lifted from a lying position, and by six months they can raise their heads when lying. Gradually too the baby uncurls from the initial birth position. Lying on their stomachs, babies soon learn to push up and support the weight of head and shoulders on the forearms. Physical control increases rapidly. By ten weeks most babies can roll from a side position on to their backs, and a month later can roll from front to back.

Growth & Development

At three months the back still sags in a sitting position, but by six months the baby can sit upright supported by cushions or in the corner of an armchair. As physical ability develops, so too the baby increasingly responds to the people around, waving arms and legs, giving and receiving pleasure from interacting with other people.

The ability to balance is a major milestone in an infant's development. At first the baby balances in a sitting position by leaning forward and placing one hand on the ground. By the age of one year, however, he or she can turn and twist around, reaching outward to grasp a toy. Few babies can crawl before the age of six months, but many are quite mobile, propelling themselves along on their stomachs by pushing with the feet. By ten months most babies can pull themselves along by their elbows and a month later the majority are crawling with the stomach off the ground.

FACES

When we look at other human bodies, we usually concentrate on the face. Facial features help us to recognize people, and their facial expressions show if they are happy or sad, friendly or angry. Our features are largely inherited, under control of the genes, which is why we resemble our parents.

BELOW Pattern recognition is a well-studied aspect of behavior in very young babies. Researchers record the times at which infants show interest in various patterns by noting reflections in their eyes. Not surprisingly, two-month-old babies repeatedly prefer the "real" face to a scrambled face or a non-face (upper diagram). Pattern is more important than color or brightness, as shown by the timings in the lower part of the diagram which demonstrate a strong preference for a face, printed letters, and a bull's eye as opposed to plain colored disks.

Growth & Development

All of these early movements are preparation for the hurdle of walking. By about 11 months many babies are pulling themselves into an upright position using a helping hand or the edge of furniture. Balance is still uncertain, however, and without support the baby falls back down again. In the next stage, the baby, still using a support, gains the confidence to shift weight slightly and take a first shuffling step to the side.

The age at which a baby takes those first solo steps varies considerably. Initially, the baby travels only short distances, from one piece of furniture to another, then increases the range until he or she can cross the room tottering unsteadily on two feet. Progress is not always continuous. Often the baby reverts to crawling, saving energy for another attempt. Illness too may cause a setback, but in general during this period the baby steadily moves toward the walking stage with all the changes that accompany this development.

CHANGING SHAPE
As a body grows from a baby to an adult, its proportions change. Compared to the overall height, the head gets smaller, while the limbs, especially the legs, become longer.

PHYSICAL AND MENTAL SKILLS

During the first year the baby gains considerable physical and mental skills. These occur because of the way in which the brain and nervous system mature during this period. At birth many of the baby's nerves are still immature. They are not coated with the insulating, fatty myelin sheath required for the complex interaction between nerves and muscles that produces voluntary physical achievement. By the end of the first year, however, myelination of the nerves is virtually complete, enabling the baby to walk and achieve other physical skills.

The brain also undergoes considerable development during the first 12 months. At birth an average baby's brain weighs 26 percent of its adult weight. But by the age of 12 months, it already weighs more than half its adult weight. By the second birthday the brain has virtually achieved maturity and grows only slightly thereafter. Such massive growth is allowed for by the membranous joints between bones of the skull, which are not finally fused together until the age of four.

By 15 months most babies are toddling happily. By 18 months most can pull themselves upright from a sitting position, and by their second birthday the average toddler is not only walking but running, dodging, stopping in mid-run, and may be sufficiently coordinated and confident to play and kick a ball without tumbling over.

Growth & Development

THE CHILDHOOD YEARS

By the age of two the helplessness of babyhood has been left far behind. Over the coming years the developing child becomes increasingly independent both physically and emotionally. Growing physical dexterity, plus a wealth of social experiences, enable the child to develop the abilities to think, reason, and adapt to new situations and people.

One major hurdle on the way to independence is scaled around the age of two when preschool children begin to learn how to control bladder and bowel movements—achievements that require physical development but which also demand the ability to understand what is expected.

During the childhood years growth is slower and steadier than during infancy or adolescence. The child becomes longer and slimmer, and body proportions change until they are nearer those of an adult. Girls tend to grow more quickly than boys, although by the age of five both sexes have reached half their adult height. From then until puberty the two sexes alternate, until by the age of 12 girls tend to be slightly the taller.

BECOMING AN ADULT

The time when someone changes from a child into an adult is called puberty. For girls, puberty usually begins between the ages of 10 and 14 years. Boys are usually 12 to 15 years old.

Height

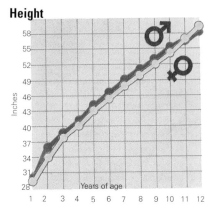

The rate of physical growth during childhood differs for boys and girls. For most of the time, boys (blue lines) are taller and heavier than girls, but at age 11 girls (yellow lines) overtake boys. The rate of growth also differs between the sexes (opposite page), with girls gaining height more rapidly than boys except in the four years between ages five and nine. All the statistics in these charts are averages taken from large samples of children; an individual child's height and weight, or the rate at which he or she grows, may be quite different from the average at one particular age or at a particular stage of development.

Weight

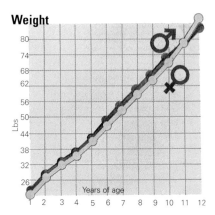

Growth & Development

Growth is controlled by hormones. The hypothalamus in the brain produces growth hormone releasing factor (GRF), which acts on the pituitary gland at the base of the brain and stimulates it to release the growth hormone somatotropin. This hormone is carried in the bloodstream to the liver and kidneys, where it is modified into somatomedin. It is this substance that finally stimulates bone growth and the calcification of cartilage, a process which hardens it to bone.

Toward the end of adolescence growth stops as the production of growth hormone inhibiting factor (GIF) rises, which critically prevents the release of somatotropin. If this does not happen, a number of disorders may result, including gigantism. Dwarfism, in contrast, is caused by failure to produce enough growth hormone during childhood.

Secondary, or permanent, teeth appear from about the age of six, the first to erupt being the back molars. Thereafter teeth appear in much the same order as the milk teeth, with the new teeth displacing the old ones. Overcrowding may be a problem and require attention if it occurs, but as with other aspects of growth and development, the course of teething depends partly on genetic factors and partly on the environment—in this instance on diet in particular.

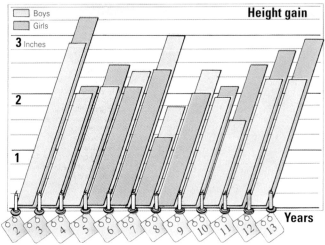

Height gain

Boys
Girls

3 Inches

2

1

Years

2 3 4 5 6 7 8 9 10 11 12 13

LATER YEARS

Our bodies are fully formed by the time we are about 20, then we slowly start to age. Gradually, our muscles become weaker and we can't move so fast, bones break more easily, joints become stiffer, skin loses its stretchiness, and hair turns white.

Growth & Development

THE TURBULENT TEENS

Adolescence or the teenage years mark the transition from childhood to adulthood. Like other stages of development, it is marked by physical and psychological changes, both closely interlinked. These can be confusing years as the adolescent struggles with self-consciousness, the drive for independence, and a search for personal identity. Family problems, too, may impinge on an adolescent's development because by this stage parents, who are no longer responsible for young children, may be making major changes to their own lives.

Puberty dominates the early teenage years. In both sexes its onset is stimulated by increasing levels of sex hormones in the bloodstream, and in girls occurs at about the age of 12, on average two years earlier than in boys.

In girls puberty is controlled by luteinizing hormone (LH) and follicle-stimulating hormone (FSH) secreted by the pituitary gland. Luteinizing hormone stimulates the ovaries to produce another hormone, progesterone, which is responsible for breast development and the onset of menstruation. Follicle-stimulating hormone brings about the release of estrogen, which initiates the development of female sexual characteristics.

The first external sign of puberty in girls is nipple development. It may begin from the age of eight, and is usually noticeable by the time the girl is eleven.

STARTING TO WEAR OUT
The main reason why our bodies start to wear out as we get older is that cells take longer to reproduce themselves. This means that it takes longer to repair and replace damaged parts of the body.

The areola, the pigmented area around the nipple, enlarges and over the next two and a half years the breasts become fully formed. Pubic hair also begins to grow, vaginal secretions increase and about two years later the menarche, or first menstrual period, occurs. Regular ovulation usually begins about a year later.

Puberty in boys usually begins between the ages of 13 and 15½. As in girls, the onset of puberty is controlled by hormones released by the pituitary gland. Testosterone, the male sex hormone, is produced by the testes and is responsible for sexual development. The testes and penis increase in size, the scrotum grows, and pubic hair develops. Facial hair also develops and the boy's voice "breaks" or deepens as the larynx increases in size.

THE ADULT YEARS

Strange as it may seem the aging process begins with late adolescence, but it is not noticeable until much later on. For most people the twenties and thirties are a time of peak physical fitness. Growth is complete and there are very few age-specific disorders to contend with. Instead health is closely linked to lifestyle: regular

Growth & Development

exercise, a balanced diet, and abstinence from cigarette smoking and excessive alcohol consumption prevent serious and life-threatening problems such as cancer and heart disease later in life.

The turbulence of adolescence has now passed and most young adults are concerned with making choices about careers, relationships, and lifestyle. The conventional path toward marriage and children is changing. Statistics today show that people in the West are choosing to marry later or to remain single for life. And surveys conducted during the mid-1980s showed that more than half of American households consisted of only one or two people. Thus the traditional two-parent family is now in a minority.

BELOW One of the penalties of aging is the inevitable wear and tear on the larger weight-bearing joints, such as the hips, knees, and spine. The result may be the crippling disorder osteoarthritis, in which the smooth layer of cartilage that normally lines and lubricates the joint gradually erodes away. An excellent remedy, particularly for the hip joint, is an artificial replacement. The head of the thigh bone, or femur, is removed and replaced with one made of metal or metal and plastic. Such joint-replacement operations have done much to restore mobility and improve the quality of life for thousands of elderly people.

THE AVERAGE LIFESPAN
Because of medical discoveries and improvements in living conditions, people now live longer than before: in 1815 people lived an average of 39 years, 1900—46 years, 1940—60 years, 1961—71 years and in 1990—75 years.

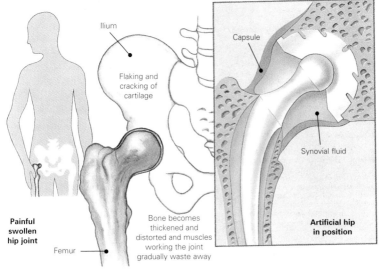

Ilium

Flaking and cracking of cartilage

Capsule

Synovial fluid

Painful swollen hip joint

Femur

Bone becomes thickened and distorted and muscles working the joint gradually waste away

Artificial hip in position

Growth & Development

THE MIDDLE YEARS

Physical and emotional changes begin again during the middle years. For many people middle age is a time of crisis, often comparable with the upheaval of adolescence. Children may be leaving home, there may be worries about sexuality, and there is an increasing anxiety about the onset of old age.

For women the middle years are marked by the menopause, the ending of their reproductive period. It usually occurs between the ages of 45 and 55, its onset coinciding with a decrease in the levels of the sex hormones. Menstruation and ovulation cease, and fertility comes to an end.

The menopause and post-menopausal period is usually accompanied by various physical symptoms, such as hot flushes, and physical changes such as vaginal atrophy and osteoporosis (weakening of the bones). Hormone replacement therapy (HRT) may ease these symptoms, but it is a controversial form of treatment because of a possible link with uterine and breast cancer.

BODY SHAPES
Our final body shape is often determined by inherited factors, there are three body shape categories (somatotypes): ectomorphic—lean and low in fat, angular and delicate appearance, mesomorphic— evenly proportioned, with strong muscular frame, endomorphic—more rounded, with a propensity for fat storage.

Later middle age may also be marked by various social changes, most especially retirement. For some retirement is a welcomed rest from the rigors of work, for others it is a distressing experience involving separation from colleagues and lifelong routines. Because of the modern trend toward younger retirement, many people finishing work are still physically and emotionally capable of making much of the retirement years. Retirement is a time of new opportunities best faced with thorough and sufficient preparation.

THE LATER YEARS

It is difficult to say when old age begins, particularly today when life expectancy has increased by more than 25 years over the last century. Even so, the years from 55 onward are marked by the obvious signs of the aging process.

Most changes are gradual. As a person gets older the skin loses elasticity and subcutaneous fat disappears, causing wrinkling and sagging. Pigmentation also changes and hair loses its color. Muscles shrink as they are replaced with fibrous tissue, a process that begins in the thirties and, for women, accelerates after the menopause. It is thought to be caused by impaired calcium absorption and calcium loss. The bones become brittle and are more liable to fracture. Cartilage thickens and joints become stiffer.

Disease and debility, however, are not inevitable in old age. Normal aging does involve a decline in physical and mental

Growth & Development

ability, but research shows that bad habits such as smoking, a sedentary lifestyle, and poor diet are major contributory factors in disease in the later years. Switching to a healthy lifestyle can do a great deal to combat the adverse effects of aging.

Aging itself is a gradual process, and decline is most usually caused by illness rather than age in itself. Today the main concern of gerontology, the study of aging processes, is to find a way of providing remedies and treatments for the common disorders of aging, and to discover a means to postpone old age or extend longevity. Much work has been done in this area, but for the moment it is a matter of luck—notably in the way you choose your parents—and healthy lifestyle.

RIGHT Wrinkled skin and gray hair typify the aging process. Other noticeable changes include a loss of height due to a shortening of the spine, often exaggerated by a slight stoop. Bones become more porous and brittle, and the senses of hearing, taste, and smell become less acute.

MALE BALDNESS
Male pattern baldness is typified by a characteristic loss of hair from the temple area first, creating a receding hairline, and often leaving hair growth at the sides of the head only. Hair loss is often age related, and allied to hormonal changes in the body as hormone production declines.

Graying of hair

Hearing less acute

Reduced ability of the eyes to focus

Taste and smell diminished

Loss of height on average approximately 3 in. (7.6 cm), and possibly a stoop

Loss of tissue elasticity causes skin to wrinkle

Steady exercise delays muscle fiber loss and maintains strength

Joints and bones become troublesome. Thinning of bones causes them to become lighter and more brittle

Limit for hard work lowered

Growth & Development

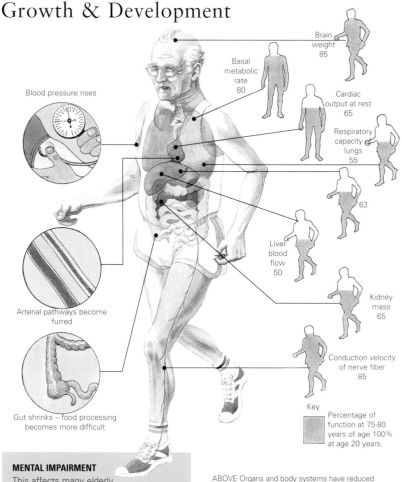

Brain
weight
85

Basal
metabolic
rate
80

Blood pressure rises

Cardiac
output at rest
65

Respiratory
capacity of
lungs
55

63

Liver
blood
flow
50

Kidney
mass
65

Arterial pathways become
furred

Conduction velocity
of nerve fiber
85

Gut shrinks – food processing
becomes more difficult

Key

Percentage of
function at 75-80
years of age 100%
at age 20 years.

MENTAL IMPAIRMENT
This affects many elderly
persons. Conditions such as Alzheimer's
disease and stroke result in reduced
mental ability, other diseases include
fracture, heart attack, and incontinence.

ABOVE Organs and body systems have reduced
capacities in the elderly. Assuming that they work at 100
percent capacity at age 20, the diagram indicates the
percentages for somebody in his or her late seventies.
Most affected are the waste-processing ability of the
liver and kidneys, the respiratory capacity of the lungs,
and the pumping efficiency of the heart. Nerve
conduction also slows down, and brain weight declines.

GENETICS & HEREDITY

Each human being is unique—and yet at the same time a person is the sum total of characteristics inherited from each parent and passed on from parent to offspring. The process by which this happens and the selection of particular characteristics is the result partly of precise programing and partly of random selection.

Heredity is the mechanism by which characteristics are passed from one generation to the next. Inherited characteristics include everything from the obvious features such as hair color and physique to less obvious blood grouping, metabolism, and enzyme grouping that determine body chemistry.

The detailed study or science of the mechanisms of heredity, and of variations that can occur both in animals and in plants, is called genetics. During the twentieth century this science has blossomed, enabling us not only to understand why we inherit certain characteristics, but also to pinpoint and treat a variety of serious human disorders—and even, however controversially, to intervene in the creation of life itself.

RULES OF INHERITANCE

Heredity is controlled by the genes. Half of each person's genes come from the mother, half from the father. Like begets like and, as a result of genetic inheritance, offspring almost invariably resemble their parents. But there are also variations. Although the basic genetic components of heredity are passed equally, they combine themselves in an almost infinite number

of ways so that there is very little likelihood of any two people—except identical twins—having exactly the same genetic makeup.

Random chance plays a large part in heredity, but there are also certain rules and patterns. The first of these were uncovered by the Moravian monk Gregor Mendel during the nineteenth century in a series of now famous experiments with garden peas. As a result of Mendel's work combined with subsequent knowledge of cell structure we know that inherited traits are passed on through the genes, and that each person has two copies of each gene, one copy inherited from the mother and one from the father. When the genes are passed from parent to offspring, either of the mother's copies or of the father's may be passed. By random mixing, a vast number of conformations

SEX CELLS

Each male and female sex cell (sperm and ovum) has 23 single chromosomes. When fertilization takes place, these 23 single chromosomes pair up, creating 23 pairs of chromosomes. These pairs of chromosomes are then copied into every cell that forms the new embryo.

Genetics & Heredity

is possible.

The handover of genetic information from parents to child takes place at conception when the two cells, sperm from the father and ovum from the mother, each of which contains only half the number of necessary chromosomes, fuse together to form one cell carrying its full chromosomal complement.

Not all pairs of genes are, however, equal in effect. One may be what is known as dominant, the other may be recessive. When this happens the dominant gene suppresses the recessive one if both occur together. The recessive gene is apparent only if both genes are recessive. Red hair and very blond hair seem to be examples of recessive inheritance of "normal" features. Otherwise recessive inheritance of "normal" features is uncommon.

Genes may also be codominant. The best example of this is blood grouping. There are three possible forms of the major blood group gene—A, B, and O. Any gene can have alternative forms—called alleles—but one person can have only two of them. Genes for group A and

B are codominant, so that somebody who inherits both an A and a B gene has group AB blood. The group O gene is recessive and therefore can produce group O blood only in somebody who has two O genes.

Exceptionally, however, a person with group A blood conceiving with a partner with group B blood may produce an O blood group offspring if they both also have a "silent," or recessive group O gene. It is this patterning that allows blood group genetics to help in some cases of disputed paternity.

Female

Ovum

Ovum

Male

X chromosome

X chromosome

Sperm

X chromosome

Sperm

Y chromosome

Female zygote

Male zygote

ABOVE Boy or girl? The sex of a baby is determined at the moment of conception by the sex chromosomes inherited from the parents. All eggs, or ova, produced in a woman's ovaries contain an X chromosome derived from the pair of X chromosomes in the cells that split to form them. Normal male body cells, however, contain one X and one Y chromosome, so that when cells divide in the testes to form sperm cells, half carry an X chromosome and half carry a Y chromosome. When an ovum and a sperm combine at fertilization, the chances are even that the resulting cell (zygote) will be XX (female) or XY (male).

GENETIC CODE
Everyone has about 100,000 genes, each made of a combination of nucleotide bases.

Genetics & Heredity

SEX CHROMOSOMES AND SEX LINKAGE

Genetic material is carried on the chromosomes—rodlike structures in the nucleus of every cell in the body. Each cell contains 46 chromosomes arranged in 23 pairs. The only exceptions are the sex cells, which contain only 23 unpaired chromosomes so that at conception, when sperm and egg fuse, they create one cell containing a full set of chromosomes. Of the 23 pairs produced, however, one pair—the sex chromosomes—differ from the others in that they are not always identical. In women the sex chromosome consists of two identical X chromosomes; in men the sex chromosome consists of one X chromosome and one, smaller, Y chromosome.

This difference in size is responsible for a number of sex-linked traits. Hemophilia is a blood-clotting disorder caused by the absence of a protein— factor VIII—usually carried on the X chromosome. A woman, with two X chromosomes, has two factor VIII genes, so that if one gene is abnormal, the other produces enough blood-clotting factor to prevent excessive bleeding. A man, by contrast, with only one X chromosome, cannot produce enough clotting factor if the one gene is faulty because there is no second gene to replace it. For this reason, a woman carrying one abnormal gene does not develop hemophilia herself, but can pass on the trait to her male offspring.

Sex-linked traits can skip generations, notably in the case of colorblindness. The recessive gene for colorblindness is carried on the X chromosome. A colorblind man can pass the X chromosome with the abnormal, recessive gene only to his daughters so that all his children will have normal color vision because the daughters' other normal X chromosome will mask the recessive effect of the gene, while the sons will receive their X chromosome from the mother. But all the daughters are carriers of the defective gene—so their sons may inherit the X chromosome bearing it and so be affected. Grandfathers and grandsons, but not sons, may therefore be color-blind.

RIGHT DNA's twisted ladder has sides made of alternate sugars and phosphate groups, and "rungs" consisting of pairs of nitrogen-containing bases held together by hydrogen bonds.

THE SHAPE OF DNA
DNA is a long chemical made of two corkscrew-shaped pieces intertwined like a piece of rope. The shape is called a double helix.

Genetics & Heredity

POLYGENIC INHERITANCE

Chance plays a large part in inheritance. There is no way of forecasting, for instance, whether an ovum will be fertilized by sperm containing an X chromosome to become a girl, or by a Y chromosome to become a boy. Either is possible. This occurs because chromosomes are sorted at random when packaged into eggs and sperms—a principle called "independent assortment." And it is believed that this principle applies to all genes.

Most human traits, such as height, are determined by groups of genes acting together—a pattern known as polygenic inheritance. But when this occurs, inheritance becomes complicated and virtually unpredictable. Perhaps only two genes—AA and BB—control height. But

ABOVE and RIGHT The bases of DNA always pair with the same partner: adenine (drawn in blue) always joins with thymine (orange), and guanine (red), with cytosine (green).

each gene may have two forms or alleles— A and a or B and b. (The capital notation is traditionally used to denote the dominant, the lower case the recessive.) A and B genes make a person taller, a and b genes make a person shorter.

An average sized person has one of each, producing a genetic makeup of AAaa BBbb. But when two average sized children grow up and themselves have children, the genes are sorted randomly. Each person's sex cells or gametes contain AABB, AAbb, aaBB, or aabb genes. Each is equally likely, and when they fertilize each other the chances of any one egg meeting any one sperm are equal. So, statistically speaking, an AABB egg is fertilized by an AABB sperm once in 16 times; the same is true for the genes that make people shorter than average—one child in 16 inherits all four shortness alleles.

PROBLEM GENES

Scientists today are working on ways of identifying genes that cause problems such as illness. More than 100 diseases, such as muscular dystrophy, are known to be caused by faulty genes. And many others, such as breast cancer, are linked to certain types of gene.

Genetics & Heredity

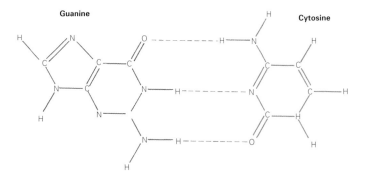

Guanine

Cytosine

As a result, so-called average parents may have children with far from average height, although average parents produce average sized children just under 50 percent of the time.

Variation is continuous. Tall parents usually have tall children, but not as tall as their parents. Similarly, short parents often produce children who are taller than they are. This is known as "regression to the mean" and it also operates in many other human traits, such as intelligence. Although heredity is random, the laws of statistical probability are used by geneticists to investigate particular traits and in genetic counselling.

A DNA MOLECULE
A typical DNA molecule is so long and thin that if it were the thickness of spaghetti, it would be 5 miles (8 km) long.

CHANGEABLE GENES

All genes can change or mutate as a result of various environmental factors, the most talked about today probably being ionizing radiation. Most mutations are harmful, in that they usually result in the death of the mutated organism even before it is born, although some useful mutations have been used in selective crop and animal breeding to produce specific strains. Nevertheless, mutations provide an essential trigger for evolutionary change, even among human populations.

In human mutations and mutagens—the substances that cause mutation—are extremely significant. If mutations occur in the sex cells they are not apparent to the person carrying them but are likely to be inherited by the next generation. It has been argued, for instance, that Agent orange, a chemical defoliant used during the Vietnam War, may have caused

Genetics & Heredity

damage to the sex cells of people exposed to it. The claim that deformed offspring resulted is still under review. The same argument also rages over the descendants of those exposed to radioactive fallout after the bombing of Hiroshima and Nagasaki, and of those members of the military who were required to witness early nuclear test explosions.

Most mutations, however, occur in somatic or non-sex cells. These may not affect heredity, but can nevertheless cause dramatic changes. Body cells are constantly replacing and replicating themselves, so that if mutation takes place it is carried through to the new cells. As with other genes, mutant ones may be dominant or recessive. Dominant mutations probably affect any offspring, although recessive mutations may not become apparent until the carrier produces children with somebody carrying a similar recessive gene. The odds of this occurring are many thousands to one except in the case of close relatives—particularly first cousins, who have grandparents in common.

Cancer is now thought to be caused by genetic damage to cells. Instead of being programed correctly, cancer cells are mutated ones with the capacity for unlimited growth and replication; they can also spread throughout the body. Agents that cause cancer—carcinogens—include chemicals, radio activity, tobacco, asbestos, and

ultraviolet rays from the sun. The effects of radiation are the subject of considerable debate. As a result of recent studies it is now known that ionizing radiation does increase the risk of certain cancers, particularly leukemia (cancer of the blood cells), and that this effect can

EYE COLOR
Most characteristics result from a mixture of two sets of genetic instructions contained in one or more gene pairs, but some features, such as eye color, are determined by a single gene. This dominant gene overrides the instructions of the other recessive gene.

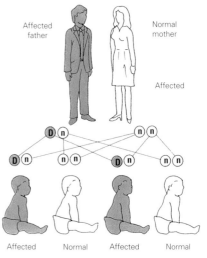

Dominant inheritance

Affected father

Normal mother

Affected

Affected   Normal   Affected   Normal

Genetics & Heredity

be produced by very low doses such as a medical X-ray of a pregnant woman. As a result there is much debate today as to whether there is, in fact, any "safe" dose of radiation.

Recessive inheritance

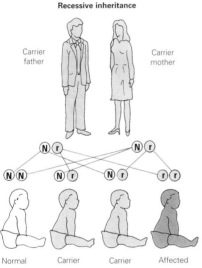

Carrier father

Carrier mother

Normal   Carrier   Carrier   Affected

LEFT and ABOVE A gene reveals its presence, or fails to do so, depending on whether it is dominant or recessive, present alone, or as a similar pair. A characteristic determined by a dominant gene (left) manifests itself even if present only singly. A recessive gene, on the other hand, has to be part of a pair to reveal its presence. A person with a single recessive gene is a carrier, who does not display the characteristic but passes it on, half to his or her offspring. In the diagram (above) both parents are carriers and so one child in four will inherit a pair of recessive genes and reveal the characteristic.

DNA AND THE GENETIC CODE

Cell division is the basis of heredity. Each cell in the body contains within its nucleus the information that controls the minute-by-minute activity of each cell and of the whole body. It also contains all the genetic information passed from parent to offspring through the reproductive process. This information is encoded in the form of a substance called deoxyribonucleic acid, or DNA. This remarkable biological chemical has two essential characteristics: it can store information and make an exact copy of itself.

The DNA molecule is a giant in chemical terms but even so is much too small to be seen with even the most powerful optical microscope. Its shape resembles that of a twisted ladder—the famous double helix—with millions of rungs. The "sides" of the ladder are composed of alternating units of phosphate and a sugar (deoxyribose); the "rungs" consist of a linked pair of chemical compounds called nucleic acid bases.

MAKING COPIES
During normal cell division and duplication by mitosis, the two strands of the DNA molecule are pulled apart, rather like being unzipped down the middle. A replica is assembled on each strand, corresponding exactly to the missing one.

Genetics & Heredity

There are four bases, designated adenine (A), thymine (T), cytosine (C), and guanine (G). A can link only with T; C links with G.

At cell division DNA can be seen because it is located on the chromosomes which become visible at this time—and genes are short segments in the threads of chromosomes. During the normal process of mitotic cell division, by means of which cells multiply, chromosomes first duplicate themselves and are then pulled apart during the splitting of the "mother" cells so that each of the new "daughter" cells gets exactly one complete set.

But there is one exception to this normal process. This occurs when the male and female sex cells, or gametes, are formed. The purpose of division in this case is to produce a cell containing only half the normal complement of chromosomes. Known as meiosis, this type of cell division is accomplished by the chromosomes splitting twice, first to separate the duplicate chromosomes into separate cells, and secondly to separate the chromosomes pairs into the daughter cells.

MUTATION

Most of the time DNA is faithfully copied so that identical genes are formed, but a small mistake in the sequence of bases results in the wrong instructions being sent out and the wrong protein being made in the factory. This alteration of the code is the basis of mutation.

Because there are 23 pairs of chromosomes in humans, there are more than ten million different ways that the chromosomes can become sorted during meiosis. In a fertilized egg, each chromosomal pair is made up from halves from each parent, consequently the chances of children of the same parents getting exactly the same set of chromosomes are so infinitesimally small that it effectively never happens. Only identical twins—from a single fertilized egg—can have identical chromosomes.

THE LANGUAGE OF LIFE

All living things contain DNA. Between the simple circle of the DNA is a bacterium and the tightly coiled spirals of human chromosomes. DNA runs in an everlasting thread of life on earth. The model of DNA structure produced by Watson and Crick stemmed from research into work on the identification of bacteria begun at the time of the introduction of antibiotics to combat bacterial disease during World War II. Bacteria were ideally useful to geneticists because they reproduce so quickly—in minutes rather than days, months, or years. Other laboratory experiments using fungi, which also grow and multiply rapidly, showed that individual genes controlled the production of specific amino acids, the building-blocks of protein, inside a cell.

DNA works all the time. To use computer jargon: during reproduction DNA is arranged into disks and files

Genetics & Heredity

(chromosomes and genes) outside the nucleus, and these transfer the master program (including the secret of reproduction) to the new cell. At all other times DNA acts as the cell's operating system, residing in so-far undetermined form inside the nucleus, and constantly managing the correct sequencing and assembly of amino acids in the protein factory of the cytoplasm.

SELF-REPAIRING MECHANISM

The library contained within the double helix of DNA includes the coded templates for all the proteins and their subsequent organization within the cells of the human body. It also holds the formulae for all the many enzymes which regulate the cell's activity: forming the ribosome RNA into a copy of a gene, identifying the sequence of amino acids, and stitching them into protein chains.

There is even a set of enzymes which police DNA itself, clipping out damaged nucleotides or filling in gaps so that the original sequence is restored. Each cell has at least 50 of these DNA maintenance enzymes on call at all times.

INHERITANCE OF HUMAN BLOOD GROUP

Possible Gene Combinations	Result in	Blood Group
$I^A I^A$ or $I^A I^O$	Antigen A only	A
$I^B I^B$ or $I^B I^O$	Antigen B only	B
$I^A I^B$	Antigens A and B	AB
$I^O I^O$	No antigens	O

ABOVE The six possible combinations of the three genes for blood type result in the four common blood groups; IA and IB are dominant (producing the A, B, and AB groups) over the recessive IO.

EVOLUTION

Nature constantly improves itself; not through any great master plan, but through constant trial and error. If a mutation is beneficial to the species, it is likely to succeed; if it is harmful, then the creature dies and the altered DNA disappears without trace. Without constant, random mutation creating improvement and diversity life on earth would cease to evolve, and a billion-year process would stop.

Glossary

Achilles' tendon the strongest tendon in the body.

adrenal gland one of two glands, each above a kidney, that secrete hormones, epinephrine (adrenaline), and other substances.

alveolus one of the thousands of tiny air sacs in the lungs where blood exchanges carbon dioxide for oxygen.

amino acid one of the chemicals from which a protein is built.

androgen a general name for a hormone that produces masculine characteristics, such as testosterone.

aorta the principal artery in the body.

apocrine gland a gland that contributes part of its own cellular substance to its secretion, such as certain axillary and genital sweat glands.

areola the lightly pigmented ring that surrounds the nipple of the breast.

artery a blood vessel that carries blood away from the heart.

atrium one of two upper chambers of the heart.

axon the highly elongated extension of a nerve cell which conducts the nerve impulses.

basal ganglion a cluster of neurons at the base of the brain.

basophil a type of white blood cell.

blastocyst an early stage in the development of an embryo.

blood cells various types of cells within the blood, includes red and white cells.

brainstem the bottom section of the brain that joins on to the spinal cord.

bronchiole tiny air-conducting tubes branching from the bronchi and terminating in alveoli.

bronchus a tube that carries air into and out of the lungs.

capillaries the smallest blood vessels.

cardiac muscle the involuntary muscle of which the heart is composed.

carotid artery a large blood vessel in the neck that supplies blood.

carpus one of the bones of the wrist.

cerebellum an oval-shaped portion of the brain.

cerebral hemispheres the two halves of the cerebrum.

cerebrum the two hemispheres of the forebrain.

cervix the neck of the uterus.

chromosome DNA-containing structure found in every cell.

cilia microscopical hairlike processes of many cell types.

clavicle the collarbone

clitoris a small oval of erectile tissue at the head of the vulva.

coccyx the "tailbone."

cochlea the part of the inner ear which contains the sensory organs for hearing.

collagen a tough, elastic protein found in bone, skin, and all other connective tissue.

colon the part of the large intestine between the cecum and the rectum.

connective tissue the body tissue that surrounds, supports, separates, and protects the various body organs and structures.

coronary arteries arteries ascending from the aorta.

corpus callosum a structure composed of millions of nerve fibers that links the two cerebral hemispheres.

cortex the surface or outer layer of an organ, as opposed to the inner medulla.

cranium the vault of the skull, consisting of eight bones.

cytoplasm the parts of a cell outside the nucleus.

dendrite a narrow projection of a nerve cell.

dermis the layer of tissue beneath the epidermis.

diaphragm the muscle that permits breathing by contracting and relaxing during respiration.

DNA (deoxyribonucleic acid) the basic genetic material that is passed from generation to generation in the genes.

duodenum the first part of the small intestine between the stomach and the jejunum.

dura mater the outermost of the three membranes that cover and protect the spinal column and brain.

embryo the developing baby from the time of fertilization up to about 3 months.

endocrine glands glands that secrete substances directly into the bloodstream rather than through ducts.

endometrium the membrane forming the lining of the uterus.

endorphin a substance released by the brain.

epidermis the outer layer of the skin.

epiglottis a thin flap of cartilage that covers the entrance to the larynx during swallowing.

erythrocyte a red blood cell.

esophagus the gullet.

estrogen one of a group of hormones which gives rise to female sexual characteristics and affects female reproductive activity.

exocrine gland any gland that secretes substances into the body.

fallopian tube one of two tubes located in the female abdomen that leads from close to each ovary into the uterus.

femur the anatomical name for the thigh bone.

fetus the baby developing in the uterus from the time that organs begin to form until birth.

fibrin the protein that forms the essential part of a blood clot.

fibula the smaller of the two bones in the lower leg.

follicle a small sac or cavity, such as the hair follicle from which a hair grows.

forebrain the front part of the brain that develops into the cerebrum.

gallbladder a sac located just below the liver.

ganglion a cluster of neurons, generally of those of the peripheral nervous system.

gene the smallest factor responsible for passing the inherited characteristics from parents to offspring.

glucose a simple sugar produced as the end product of starch and carbohydrate digestion.

growth hormone, or somatropin. A chemical produced by the pituitary gland which regulates the growth of bones and other tissue.

hemoglobin the iron-containing molecule of red blood cells responsible for oxygen and carbon dioxide transport.

hepatic of the liver.

hepatic portal vein vein through which blood from the capillaries of the spleen, pancreas, stomach, and intestine are taken via the liver.

hormones secretions of ductless (endocrine) glands within the body which may affect local or distant parts of the body.

humerus the long bone of the upper arm.

hypothalamus a vital area in the brain located beneath the thalamus.

ileum part of the small intestine between the jejunum and the cecum.

inner ear collective name for the semicircular canals and the cochlea that deal respectively with balance and hearing.

insulin a hormone produced by the pancreas that regulates sugar levels in the body.

ion an atom or molecule that carries an electric charge.

jejunum the part of the small intestine between the duodenum and the ileum.

jugular vein one of four veins that conveys blood from the head and neck to the heart.

lacrimal gland a spongelike organ in the eye that secretes tears.

large intestine the continuation of the small intestine, where the final processes of digestion are completed.

larynx the voice-box.

leukocyte a white blood cell.

ligament a band of fibrous tissue that connects bone or cartilage and supports joints.

lymph a transparent, watery liquid that circulates through the body in a system of tiny vessels, in a manner similar to blood.

lymphatic system a system of glands and vessels which drain fluid from the tissues of the body together with dead cells and bacteria.

lymphocyte a type of white blood cell that plays a key role in the body's defense system.

mandible the lower jawbone.

maxilla one of the two main bones of the face that constitute the upper jaw.

medulla the central core, or inner layer of an organ.

meiosis the type of cell division taking place during the formation of sex cells.

melanin the dark pigment present in skin and hair.

meninges the membranes which form an envelope around the brain and spinal cord.

metacarpals a group of five bones that extend from the wrist to form the palm of the hand.

metatarsals five bones that articulate with those of the ankle and toes to form the arch of the foot.

midbrain the uppermost part of the brainstem.

middle ear the part of the ear that contains the eardrum and ossicles.

mitosis the division of a cell nucleus into two identical nuclei.

motor nerve a nerve carrying information from brain or spinal cord to a muscle.

mucus a slimy secretion that lubricates body linings.

myelin the fatty substance that makes up much of the sheath round many nerve fibers.

myofibril one of the many slender fibrils, consisting mainly of protein, that fill a muscle fiber.

neurotransmitter chemical substance that transfers impulses between neurons.

neutrophil the most common white blood cell.

node of Ranvier one of the gaps in the insulating myelin sheath around a nerve.

nucleolus structure within the nucleus of the cell.

nucleus the part of the cell which contains the genetic material in the form of DNA.

olfactory concerning the sense of smell.

olfactory nerve the nerve that extends from the top of the nose to the center of the brain.

optic chiasma the junction of the two optic nerves.

osmosis The diffusion of a solvent through a semipermeable (porous) membrane from a weak to a strong solution.

ossification hardening.

ovary one of a pair of glands in the female.

ovulation the shedding of a mature ovum (egg) from the ovary.

ovum egg; the female reproductive cell.

oxyhemoglobin the molecule formed by the binding of oxygen to hemoglobin.

oxytocin the hormone produced by the pituitary gland that stimulates contractions of the uterus and release of milk at the breasts.

pancreas a large digestive gland which opens into the small intestine and contains tissue which produces the hormone insulin.

papilla 1. a tiny projection that presses from the dermis into the epidermis, nourishing a hair follicle. 2. one of the four types of minute elevation that cover the tongue, three of which contain taste buds.

parathormone a hormone produced by the parathyroid glands.

parathyroid gland one of four endocrine glands in the neck.

parotid gland one of two saliva-producing glands.

pepsin an enzyme produced in the stomach that assists in the digestion of protein.

peritoneum the lining membrane that covers

the organs contained within the abdominal cavity.

phagocyte any cell where function is to absorb and neutralize foreign bodies.

phalanges the bones in the toes and fingers.

pia mater the innermost of the three meninges.

pituitary gland a small endocrine gland found at the base of the brain.

plasma the clear fluid content of blood and lymph.

platelet a cell fragment found in the blood that plays a major role in clotting.

plexus a network of interwoven nerves.

pons a part of the brain stem.

portal vein carries blood from the intestines to the liver.

progesterone a hormone produced in the ovary.

prolactin a hormone that encourages growth of breast tissue and the release of milk.

prostaglandins fatty acid derivatives.

prostate gland a gland surrounding the urethra at the neck of the bladder.

puberty the stage when an individual matures sexually.

pulmonary of the lungs.

radius the shorter of the two lower arm bones.

rectum the lower part of the alimentary canal.

red blood cell a disk-shaped blood cell containing hemoglobin.

reflex arc the path traveled by a reflex.

renin an enzyme secreted by the kidneys.

respiration the process of breathing in and out.

respiratory center the regulator of breathing located in the brain.

RNA ribonucleic acid, a long-molecule compound found in all cells.

sacrum a triangular bone, consisting of five fused vertebrae, that forms a wedge between the two hip bones.

scapula the shoulder blade.

scrotum the external muscular sac surrounding the testes.

sebaceous gland a skin gland associated with hair follicles.

semen a sticky white fluid made by the male reproductive organs that carries sperm.

seminiferous tubule any of the small tubes of the testicles that produce sperm.

serum the clear component of blood which separates into liquid and solid elements.

sex hormone any of the chemical controllers of sexual characteristics and reproduction.

skull a skeletal case enclosing and protecting the brain.

small intestine, or ileum. A tube in which most of the processes of digestion take place.

sperm the male reproductive cell.

sphincter a circular muscle that controls the opening and closing of a hollow organ, such as the heart.

spinal cord nerve tissue contained within the vertebral canal in the backbone.

spinal nerve one of the nerve trunks that emerges from the spinal cord.

spine the chain of

vertebrae running from the cranium to the coccyx.

spleen an organ situated in the abdominal cavity near the liver.

sternum the breastbone; a daggerlike structure to which most of the ribs are attached.

synapse the "gap" or region where two nerve cells come into close contact.

synovial fluid a lubricating substance secreted by the membrane of a joint.

talus an ankle bone.

tarsals seven short bones that make up the ankle.

tendon a fibrous cord of connective tissue that attaches muscle to bone.

testicles the two reproductive glands that produce male reproductive cells and testosterone.

testosterone one of the sex hormones produced in men.

thalmus the relay center of the brain from where sensory impulses are directed to other brain areas.

thymus gland a ductless gland of the lower neck region which disappears by the time of adulthood.

thyroid gland a gland in the neck.

tibia the shinbone.

trachea the windpipe.

ulna the longer of the two bones of the forearm.

ureter a tube extending from the kidney along which urine passes to the bladder.

urethra the duct by which urine passes from bladder to the exterior.

urinary tract both the

passage and the organs connected with the production and discharge of urine.

urine a watery solution of waste products removed from the blood by the kidneys.

uterus the womb.

uvula the visible projection suspended from the midpoint of the soft palate arch at the back of the mouth.

vagus nerve a nerve that controls intestinal movement.

vas deferens the excretory duct of each testicle that conveys sperm.

vein a vessel that carries deoxygenated blood toward the heart.

vena cava either of the two major veins that empty deoxygenated blood into the right atrium of the heart.

ventricles the two lower chambers of the heart.

vertebra one of the 33 bones that form the spinal column.

villus a small protrusion from the surface of a membrane.

white blood cell any of several types of blood cell that have a nuclei.

X chromosome a sex chromosome. Unlike the Y chromosome it carries major genes, which show sex-linkage. Females have two X chromosomes.

Y chromosome a sex chromosome. Males have one Y chromosome and one X chromosome.

zygote the fertilized ovum.

Index